# A Complete Book on Harvest Technology and Horticultural Crops

A Complete Book on
# Harvest Technology and Horticultural Crops

Tek Bahadur Subba

**CYBER TECH PUBLICATIONS**
4264/3, Ansari Road, Daryaganj, New Delhi-110002 (India)
Ph.: 011-23244078, 011-43559448 Fax: 011-23280028
E-Mail: cyberpublicationsdelhi@yahoo.com
Website: www.cybertechpublications.com

A COMPLETE BOOK ON HARVEST TECHNOLOGY AND HORTICULTURAL CROPS

Tek Bahadur Subba



First Edition 2010



*Published by* : Asian Offset Printers

**G.S. Rawat** for **Cyber Tech Publications**
4264/3, Ansari Road, Daryaganj, New Delhi-110002 (India)
Ph.: 011-23244078, 011-43559448 Fax: 011-23280028
E-Mail: cyberpublicationsdelhi@yahoo.com
Website: www.cybertechpublications.com

# Preface

India is renowned for its rich diversity in horticultural crops. Though India ranks first as producer of fruits and second in vegetables in the world, our share in world market is not significant. Therefore strategic measures have to be taken to reduce post harvest losses of agricultural commodities produced in developing countries like India and to boost the processing sector to provide better returns to farmers. Post harvest technology assumes great attention during recent years since preservation of agricultural produce is a basic necessity to sustain agricultural production. It helps to add value of produce, thus having great scope for employment generation at the production catchments.

There is growing interest in all areas for storage and transportation of horticultural products, because the products are very perishable and because of the long distance between countries and their major export market. Reducing post harvest losses is very important, ensuring that sufficient food, both in quantity and in quality is available to every inhabitant in our planet. Reduction of post harvest losses reduces cost of production, trade and distribution, lowers the price for the consumer and increases the farmers income. Post harvest research has to be not only limited to storage conditions but also to market requirements, breeding and cultivation circumstances.

— *Author*

# CONTENTS

# 1
# INTRODUCTION

There is a growing interest in all areas for storage and transportation of horticultural products, because the products are very perishable and because of the long distances between countries and their major export market. Worldwide Post Harvest fruit and vegetables losses are as high as 30 to 40% and even much higher in some developing countries.

Reducing Post Harvest losses is very important; ensuring that sufficient food, both in quantity and in quality is available to every inhabitant in our planet. The prospects are also that the world population will grown from 5.7 billion inhabitants in 1995 to 8.3 billion in 2025. World production of vegetables amounted to 486 million tonnes, while that of fruits reached 392 million tonnes. Reduction of Post Harvest losses reduces cost of production, trade and distribution, lowers the price for the consumer and increases the farmers income.

***Distribution of Vegetables and Fruit Production in the Various Regions (in million tonnes)***

| | *Vegetables* | *Fruits* |
|---|---|---|
| North and Central America | 44.16 | 53.17 |
| Europe | 66.98 | |
| Europe | 66.98 | 65.17 |
| Africa | 35.20 | 54.06 |
| Asia | 292.41 | 142.24 |
| Pacific | 2.60 | 5.62 |
| South America | 16.06 | 60.95 |

**(FAO)**

| | *Production* | *Edible matter* | *protein* |
|---|---|---|---|
| Vegetables | 486 | 57 | 5 |
| Fruits | 388 | 53 | 2 |

The increase in prosperous population, the urbanization and the green revolution have enlarged the market for all kind of ornamentals. Post Harvest research has a high return.

Significant contributions, minimizing the Post Harvest losses have been made through research on the physiological changes in the product after storage, new longlife varieties, suitable the product after storage, new longlife varieties, suitable cultivation circumstances, optimum harvesting indices, recommended storage recommendations, pre-cooling refrigerated transport and careful handling.

In preserving the quality of the produce reaching the final market, the whole marketing chain from the producer to the final consumer must be examined for any weakness. The Post Harvest research topics have to be closely related to the socio-economical and financial situation, the climatical and cultivating circumstances, the market organization and the requirements of the specific commodities.

The current needs of scientific research and extension services are different between developing and high developed, industrialized countries, and the nature of the projects must be adapted, step by step, following the evolution.

In this respect, the possibilities of linkages and further international research cooperation has to be stimulated by international institutions such as EC, FAO, ISHS, HR, and through initiatives such as the World Conference on Horticultural Research.

Post Harvest research is definitely a muti-disciplinary endeavour of biology chemistry, physics, pathology, biochemistry, physiology, marketing, financing, psychology and ethics in a social-economical, political and financial context.

## Suitable Post Harvest Research in Developing Countries

Pole Post Harvest research in developing countries: Post Harvest research has to be not only limited to storage conditions but also to market requirements, breeding and cultivating circumstances.

Treatments to arrest deterioration, harvesting at correct maturity, pre-cooling to reduce field air, care during sorting, packaging, transportation are also determinent Post Harvest factors.

Some methods of handling, packing and storage have been developed over years, using innovative abilities and skills, also low cost and commonly available materials and technolgies.

In other methods different new techniques and disciplines are implemented.

Market situation: Roadside markets with small shops or a type of local marketing are seen commonly at the suburbs of a city and in rural districts, mostly close to farms and selling seasonal products.

Price differences according to quality are mostly too small and don't stimulate enough quality improvement.

Urban population is exploding in developing countries, rising from 35% of the total population in 1990 to 54% in 2020. With this increasing urban population more free market and wholesale markets will be required.

Product is shipped from neighbouring areas as well as from distant ones.

Comparison with imported products stimulates increase of the quality of the regional produce. Export will make clearly show what quality of the regional produce. Export will make clearly show what quality level is needed for exported produce. Continuously following consumers and traders requirements, growers have to be informed in the choice of the varieties, the cultivating techniques and the extending of the selling period. The presence of growers cooperatives or collecting centers for selling products and the requirements of exporters and supermarkets could stimulate for applying suitable varieties cultivation methods of sorting, grading and transporting, and establishing research activities and extensions services.

***Cultivating Conditions:*** Many small growers produce a high variability of crops, in the villages for self consumption and around the cities for selling the rest. Changing the choice of the varieties and the cultivating techniques is the most important and must be oriented to a higher quality and healthy product, preferred by consumers and with a longer shelf-life required by traders. Pest control with plant extracts has to be improved.

New developments presume proper financing possibilities, not always easy to get. Exceptionally some big specialized growers produce a high quantity of nice products for export as Phaseolus beans in Senegal or mangos in the Philippines. Grading and sorting:

***Quality Standardisation:*** Grading and sorting in quality classes is limited or not existent and this evolution can be stimulated by the development of cooperative selling organisations.

***Packaging:*** Packaging is dependant on the disposal of packing material in the country (bags, wooden crates, woven baskets). Adaptations are necessary to reduce bruising by filling and transportation and to increase shelf life. The evolution is to use standardised reusable plastic boxes and suitable corrigated carton boxes, after wooden crates.

***Transportation:*** Most growers have only at their disposal limited transport facilities as bicycles, three wheeled or small trucks. Some traders with bigger trucks can buy the products in the producting area and transport them to the markets in the cities, with high profit.

Due to the poor condition of the roads there is a rapid deterioration of the produce in some areas. Cooperatives could place at the disposal of the growers suitable important transport vehicles.

In developing countries a few number of exporters transport proper packed and graded product by refrigerated trucks, air of boat on commissions (± 10%) to countries far away.

## Storage and Processing

Storage technics are mostly limited Storage and processing: Storage technics are mostly limited to evaporative cooling or ventilated cooling with air in winter and cold water in summer (humypad cooling) or storage in caves. These technologies have to be optimized and complemented by proper ventilated or simple modified atmosphere systems.

Refrigerated warehouses in developing countries are located mainly in large cities and ports but very limited in the farmers area (garlic in Beijing).

Refrigerated storage rooms have to be developed in the production areas.

In developing countries the introduction of new expensive, not ozone degrading cool media is not easy.

Research on the optimum cooling techniques of different commodities, to improve shelf-life, have to be studied, as long-life cultivars applying biotechnology technics, better cultivating techniques and pest control, harvesting using objective maturity indices.

Reducing bruising during harvesting packing and transport, picking in the morning, putting the product in the shadow and increasing the speed of selling times, are key basic elements for quality.

To prolong storage time, heat treatment of fruit, curing of root-vegetables after harvest, chemical treatments of flowers, treatment with growth regulators or wax of fruit and vegetables, chemical drying of onions before harvest are important element.

To preserve fruit and vegetables in developing countries research on solar drying, laetables in developing countries research on solar drying, lactic acid fermentation (kimchi), fresh pack fruit added with the juice of tropical berries with very high Vit. C en low pH or anti-browning agents as grapefruit seeds must be stimulated.

Information to growers through extension services: The current emphasis of the Post Harvest extension services is to reduce losses, increase quality, to invest low cost suitable technology, using suitable financing. Proper information is necessary to help the sometimes poor and exploited producers with technical and marketing assistance, for more market directed crops.

It is recommended that a structural system for education and training of growers will be implemented.

## SUITABLE POST HARVEST RESEARCH

Industrialized countries, research is related to more fundamental research, and characterized by the introduction of far advanced expensive technologies and the integrataion of different scientific disciplines.

***Market Situation:*** Specialized wholesalers-exporters and retailers and supermarkets require big quantities of the same quality.

Farmers have to concentrate their production through cooperation, selling big quantities of the same quality, offering a variability in crop by remote selling, and supple same quality, offering a variability in crop by remote selling, and supply on a long period. Competitiveness is increasing, mostly with commodities from tropical countries, inducing some problems in selling final produces.

There exist a consumer preference, different from market to market with also market segments.

The relationship price-specific quality indice is studied for different markets (Northern Europe red tomatoes, Southern Europe pink).

For long storage time the elasticity supply-demand factor is important.

Export market research will make clearly show what level is needed for exported produce and will help to find new suitable markets.

Communication and information of the consumer about quality (health, nutritive) and the promotion of the product means a sophisticated knowledge and experience. Quality of food is quality of life.

## Cultivating Circumstances

Few farmers produce high quantities of an high quality product. They are very specialized in one crop and a few numbers of cultivars.

Through breeding and molecular biology techniques long-life and disease resistant varieties are created. The cultivating techniques and pest control methods are modelled and optimized. In the last time most interest is given to healthy aspects by integrated fruit production and environmental friendly fertigation.

The supply is extended through plastic covering and glasshouses, he supply is extended through plastic covering and glasshouses, also through storing of plants as strawberries and raspberries for later forcing.

Objective indices and apparatus indicating optimum harvest are generally available. The application of Post Harvest chemical treatments by drenching, dipping or fogging is decreasing for consumer healthy objectives. There is yet some interest for heat treatments and wax coatings.

***Grading and Sorting:*** Quality standardisation and determination: The EC quality categories E, I, II for sorting and size (mm) or fruit-weight are generally practised. This distinction is no more sufficient for traders because external and internal quality in one quality category is changing too much.

Selling boards make own quality blocks between one category (red and ground colour for apples, morphologic aspects for vegetables). Also some supermarkets require minimum quality norms for quality (colour, brix, titric acidity, firmness).

For controlled integrated fruit production labels for quality insurance are applied. International standardization of analytical and equipment is necessary. For controlling quality and automized sorting non-destructive methods are studied based on physical, acoustic, optical, electric impedance, X-ray, chlorophyll fluorescence and nuclear magnetic resonance technics. For export to some countries or quarantine security is required.

***Packaging:*** Standardisation of the dimensions of package (60 cm x 40 cm, 30 cm x 40 cm) and the pallets (80 cm x 120 cm, 100 cm x 120 cm) is far advanced.

After wooden boxes reusable uniform plastic boxes - sometimes pliable—are used (Europool).

Corrugated carton boxes with minimum compression and minimum 4,5% open air locks after humid air pre-cooling and long transport are available.

Modified atmosphere packaging with absoption layers for water and ethylene, and inserts are used. For expensive products, PP films with a wide range of micropores are of great interest. The presence of some pathogenic organisms need urgent research.

The disinfection of wooden boxes with fomaldehyde or azaconazole is important for long time storage in high humid circumstances.

Labels on boxes or product have to inform consumers about the origin, the quality and the welfare of the product.

Increasing attention is given to fresh product, fresh pack and tightly processed products and juices.

***Transportation:*** To meet the nutritional needs, higher amounts of food will have to be transported over greater distances and for a further extended time. Refrigerated trucks and containers are more used.

To reduce the dependence on expensive rapid air transport, CA refrigeraP> To reduce the dependence on expensive rapid air transport, CA refrigerated trucks and containers are performed. Accordingly, maximum values for impact and vibration norms are recommended for trucks on normal roads.

## Storage and Processing

On construction level there is an evolution from cheap cork or glass fibre insulations, and bitumen or paint gastightness to gastight polyurethane-metal panels.

Requirements for compressor capacity and relative humidity increase. New not ozone degrading refrigerant are used.

Compared to regular storage, modified atmosphere, controlled atmosphere, ultra low oxygen storage, quick pull down of oxygen and ethylene absorption are now suitable according to costs of investments and price.

New temperature, $CO_2$ and $O_2$ sensors for scrubbers and membrane separators are favouring monitoring and lowering risks of not proper storage. Pre-cooling with humid forced air cooling and ice bank has a lot of benefits for vegetables and berries.

For each commodity specific storage conditions are dressed. CA storage for tropical fruits, vegetables and fresh pack products is recommended. To extend the supply of berries, storage of plants is applied. During storage, chemical treatment of flower with preservatives is well developed.

In processing development of juices, purees, extracts and essences significant progress is perceptible, with new technics nces significant progress is perceptible, with new technics as microwave, membrane separations, ultrahigh temperature, aseptic package.

**Future Post Harvest Research Priorities**

- Genetic manipulation for long life, diseases and environmental-stress resistant cultivars.
- Modelling cultivating conditions for high quality and long life (avoiding root stress by heat or drought).
- Environmental friendly pest control.
- Objective determination of suitable harvesting date.
- Post Harvest treatments (heat, UV, irradiation $CO_2$, chemicals) for storage ability.
- Monitoring refrigerating systems
- Optimum storage conditions as CA storage for tropical fruits, ornamentals, planting material, fresh pack and lightly processed produce.
- Possibilities of modified atmosphere packaging (MAP), absorption layers, inserts sensors, PP films with micropores

- Prevention of possible pathogenic organisms in MAP products.
- Adaptive control of storage conditions with biological sensors.
- CA storage during transport and as quarantine measures.
- Humid forced air Pre-cooling.
- Minimum impact and vibration norms for bruising during sorting and packaging.
- Objective non destructive measuring of quality and maturity
- Environmental friendly packaging.
- Consumer and traders quality preferences in each country.
- Cost and return of the investment of Post Harvest techny.
- Cost and return of the investment of Post Harvest technologies.
- Fundamental research on senescence, ripening, respiration, ethylene effect, chilling, fermentation, superficial browning.

## INTERNATIONAL POST HARVEST RESEARCH COOPERATION

Post Harvest research has to be adapted to the specific Post Harvest problems in each countries - according to the cultivated commodities and the socio-economical, technical and financial situation. International contacts between scientists and international cooperation projects will be very important in future to accelerate the evolution in developing countries especially.

The information of existing research programmes in the world collected on occasion of the World Conference on Horticultural Research will be very helpful to determine the Post Harvest priorities for each country and to stimulate cooperation.

International organisations such as ISHS, ASHS, EC, FAO, could there after contribute considerably in extending Post Harvest networking, research contacts and cooperation and to support some priority research projects, together with the existing governmental research and extension service.

## PRODUCTION AND MARKETING DECISION

Thailand's horticultural export, both fresh fruit and vegetable, has increased over the years. Trade opportunity is largely due to increasing interests of major importing countries and Thailand's relative competitiveness over other producers. To increase its presence in regional/world markets, the industry is facing increasing pressures and unfamiliar trading practice.

In addition, the decade-old production system based on increasing yield and reducing production cost by the producers, as well as the traditional management style of exporters, are inadequate in the face of the challenges of an increasingly competitive and discriminating global trading environment. From the business point of view, the market operation needs to move toward improving management efficiency and effectiveness rather than developing new technology alone.

This paper provides a brief overview of international horticultural trade with special reference to tropical fruit, analyses of factors influencing production and trade and trends in market development of fresh produce, effect and implication of new emerging producers, and the future role and challenge for Post Harvest technology in response to changing market scenes.

## OVERVIEW OF GLOBAL FRESH FRUIT INDUSTRY

### Market Size

The total value of international horticultural export trades (fruit and vegetables) has increased from less than US$11 billion to US$20 billion, and to US$44 billion over the periods 1980-82, 1990-92 and 2000-02, respectively. This included fresh (70%) and processed products (30%), of which half were fruit juices.

The increase represents an impressive annual growth rate of international exports at an average of 8-10% in relation to a slower growth prospect of traditional agricultural commodities such as cereals, livestock products, oil seed, fats and oil. No country can provide all the fresh produce it consumes weekly the whole year, in turn creating trading opportunities among countries. Expansion of trade for vegetables differs from that for fruits. Suppliers vary significantly between vegetable and fruit categories. Despite the increase in international horticultural trades, a far greater amount of fresh produce produced in each country is still domestically consumed.

### Major Markets

Developed countries provide the largest production and import market. Between 60% and 80% of horticultural export trades are carried out among developed countries. Demand for fresh fruit is concentrated in Europe and the United States, which absorb around 60% and 11%, respectively, of international imports. The United States is the dominant player in the international trade of horticultural commodities, ranked number one as both importer and exporter accounting for about 18% of world horticultural trade.

In general, developing countries are major producers of tropical fruits (accounting for 98% of total 60.4 million tons in 2000), but their exports remain insignificant. International trade of tropical fruit represents less than 3% of total tropical fruit production and until recently most export went to other developing countries. It is important to note that tropical fruit is mostly local fruit suited to local taste. Trade data for 1999 indicate that Latin America, the Caribbean, and Asia accounted for about 75% of world trade in fresh tropical fruits and 80% were imported by developed countries. It is generally agreed that the potential for the sustainable growth of the tropical fruit industry depends on export to developed countries and on the size of the domestic market.

### Product Composition

Horticultural trade is characterized by a wide diversity of products. The product composition of horticultural exports has changed over time with an increase in the relative importance of fruit than vegetables. The more traditional varieties of fruit, namely citrus, apples, grapes, bananas, pears, pineapples and peaches, represent some 85% of world production with those of temperature climate origin (apples, grapes, pears and peaches) accounting for 42%.

These are mainly produced and consumed in the developed countries. Of the typical tropical fruit, only banana, pineapple, and mango are produced in significant quantities and these are mainly produced in developing countries and consumed domestically. The taste for exotic products is rapidly evolving in developed countries. It was in the past 10 years that exotic tropical fruits such as avocado, guava, papaya and kiwi have become mainstream trading items. Pineapple accounts for 44% of total tropical fruit trade, followed by mango at 27%, avocado at 12%, and papaya at 7%. Among fresh vegetables, the miscellaneous vegetable category had the largest share, followed by roots and tubers. Tomatoes were the most important single item in the miscellaneous category that also includes a wide range of varieties such as pumpkins, lettuce, cucumbers, garlic, chillies and peppers, cabbages, cauliflower, carrots, garlic, onions, asparagus and ginger.

## Significant Trends in Production, Marketing, and Export Development

Factors influencing trends are summarized into three broad categories as a result of changing demand/supply, market environment and supply chain management, and regulatory/legal requirements, taking the recent developments in Thailand as examples.

***Increased Interest in Tropical Fruit Trade:*** Tropical fruit is becoming increasingly available in Europe. The increasing interest in tropical fruit in the developed countries is due to increased promotional activities and better information to consumers. In addition, it is largely attributed to the fact that the food markets are now saturated, leading to intense competition, lower margins, and lower prices for food. Since 1990, many modern trade retailers have positioned their store formats and image around the fresh produce department. In order to further attract and delight the consumer, the markets are demanding less mass customizing of food items; diversified products based on consumer food consumption patterns; and safe, environment-friendly, organic product or by introducing new products.

The introduction of tropical fruit meets all these criteria. The variety and composition of produce must therefore be enhanced, so that the characteristics of tropical fruits as diverse, exotic, unique in taste, highly nutritional, have multiple uses, and based on low input production system can be fully appreciated. For Thailand, the recent worldwide increase in the popularity of Thai food also increased the demand for Thai fresh produce.

***Major Export Markets:*** According to the Food and Agriculture Organization (FAO), international consumption of tropical fruit will increase by around 40% between 1995 and 2005, equivalent to an annual growth rate of 3.5%. Despite the increase in consumption, developing countries export little tropical fruit due to short seasonality, short shelf-life of the fresh produce, high transportation cost, and increasing strict regulations in import. Preferences of tropical produce vary among countries and also vary greatly even within localized areas in one country.

Major trades in tropical fruit, until recently, remain limited either for domestic consumption in the country, or to regional markets or other developing countries with similar taste and preference. Exports to far away markets remain constrained with the exception of bananas with some 30% of total production exported, mainly to developed countries. As more produce enter into the trade and with increasing trend in global sourcing, market prospect for tropical fruit in developed countries are becoming positive.

***Off-Season Supply:*** Supply of tropical fresh produce during off-season, counter season and/or complementary trade will become the norm than the exception. Good examples are the shipment of Southern Hemisphere produce to meet consumer demand during the Northern Hemispheres' winter when the domestic supplies are low. Most trade is caused by differing levels of relative competitiveness between producers of the same or similar products during different seasons to avoid trade dispute. Thailand's export of asparagus, to EU and Japan, has more than twofold increase in values from 1998 to 2002 due to Thailand's natural climatic production advantage with the ability to produce almost year-round.

***Domestic Market:*** Export performance of horticultural trade differs widely among developing countries. It has been estimated that future rate of growth of domestic demand is much higher in developing countries than in developed countries and the prospect of high intra developing-country trade may not likely be sustained because each has increased production and expected future export growth. A large domestic market remains very important for export development, more so for fruit than vegetables. Domestic markets provide both a springboard for the growth of exports and a cushion to absorb the shocks or uncertainties of export markets. Thailand's longans and durians, two top export items each with more than US$50 million in export values per year representing an exceptionally high 30% and 15% of the country's production, respectively, still require a large domestic market to avoid price fluctuation during peak season. Thailand produces approximately 800,000 tons of durian and exports 110,000 tons mainly to China (via Hong Kong), Hong Kong, Singapore and Taiwan.

Though there is also an increasing interest for producers in the developed countries to exclusively produce tropical fruit for and sell in export markets, the global production statistics of key tropical fruit fail to indicate such a trend.

The relative low export indicates that tropical fruit is typically produced for domestic market and suited to local taste. A further observation is that in many cases the largest exporters of tropical fruit are not the top producing nations. This indicates that these countries are adopting a specialized export-oriented strategy, and this is largely due to a significant price differential in favour of exports and based on high cost production system, high capital investment, and a well-coordinated vertical supply chain.

Falling Price: This increasing interest in the expansion of market stimulates increase in supply and this has a consequent downward pressure on prices and tends to depress margins due to intense competition especially among exporters in the developing countries. Mangoes are a good example, with prices in Europe falling 30% since 1988, in conjunction with a 66% increase in import volumes. The increase in export market and in suppliers resulted to reduced profit margins, which discouraged many exporters, and to high cost of production. This in turn has led to a supply of relatively poor-quality produce by low-cost producers.

**Concentration of Suppliers and Their Composition**

The size of the tropical fruit export market relative to the output in many cases is not large enough to allow many countries/exporters to succeed and will initially result in a smaller share of the market held by each country/exporter. Thus, the enlargement in trade may eventually result in consolidation and concentration of fewer suppliers operating at a reduced profit margin offering produce at a comprised quality of larger volume. This is usually considered transitional when the small industry is at a crossroad. There is a general interrelation indicating that an increase in export trade often results in overall improvement in quality, but produce for sale in the domestic markets are often of lower quality or are export rejects. Product differentiation or specialization to meet the large sourcing need of modern trade may result in reduced level of overall country production though there may be increased export by individual firm. The long-term effect of such trend to the industry will need to be seriously assessed.

Farm to Table: Production methods for domestic consumption are simple, much less demanding, and the domestic market usually pays well when lower expenditure on cost structures and investment are taken into consideration. To meet the discerning export market requirements, farming practices and marketing system will have to change.

However, changes in the production methods, better sanitation and hygienic farm practice, and improvement in quality will not ensure that export market can be developed and maintained. Traditional fragmented small-sized producers and labour-intensive nature of tropical fruit production system needs to be optimized and to link to companies that organize the distribution of products in the export market.

A developing country that wants to export a not yet established tropical produce often exports it through an agent in the importing country who handles the distribution and marketing without any financial risk but for a commission. The export bears the entire risk. The risk is shared by the importer only when trust has developed over time between the importer and the exporter. Commission sales play an important role in fresh produce markets for new exporters and remain so even for established exporters. Regulations in some countries that prohibit commission sales and require that horticultural imports must have a prearranged buyer tend to inhibit fresh produce trade. There is no organized system for market intelligence and trade regulations for horticultural products comparable to what exist for other agricultural products. The wide variety and small trade involved with fresh produce do not attract adequate attention for the international or national trade intelligence and statistical services. Most traders and government in the developing countries lack the leadership and vision and are ill-prepared to formulate a well-developed sustainable export strategy.

## Food Safety Issues

The traditional approach to food safety was that people at each stage of the food supply chain were responsible for the handling of the food and, hence, for food safety issues. This is easily said than done as the food chains have become more complex and the initiation of food safety strategy is often difficult, if not impossible, in the fresh produce supply chain where there is no certain predominant market power, or worst, a consistently shifting market power as affected by supply and demand during the season.

Recent legislation trend in many developed countries is to put the responsibility for food safety into the hands of the retailer. This is an over-simplified approach to the complex food safety issues and the tendency is toward strict requirements for imports through an expensive time-consuming import/export inspection. There has been increasing pressure and requirement to implement third party certified food safety and quality management systems especially for import/export trade. While Good Agricultural Practices, Good Manufacturing Practices, Hazard Analysis Critical Control Point, and ISO 9000 certification have been focal points for many organizations across the world to be used for food safety and management of value chain, there have been no appropriate frameworks for these standards for fresh produce. Some of these systems are either unrealistic and unworkable because of inherent weaknesses/uncertainty/dynamic nature of the fresh produce industry and the difficulty to provide branding and to understand certification, or too costly because such systems are still based on inspection and or regulation, or the intended holistic approaches are not meeting customer/business specifications or formalized dialogue in the fresh produce supply chain, or due to the fact that users and the bureaucracy take a simple concept and make it difficult for operators.

Without a clear purpose, too stringent/improper safety requirements and the application of different health standards for domestic and imported goods, and imposing food safety requirements often had an obvious protectionist component. Many safety factors/variables need to be considered in harmonizing national standards with the guidelines through international consensus, and mutual recognition agreement on food safety issues is expected to take time. The likelihood is high to forge increased bilateral and regional trading arrangement. Another important consideration is the need for a dialogue between policy makers and the producers/operators. A recent bilateral agreement on fresh produce trade between Thailand and China, and the Thai government initiative on food safety project have been mistakenly interpreted as restricting trade by the producers/operators. The challenges are to provide better communications and to put in place food safety systems that are viable and effective both in terms of cost and efficiency.

## Trade Liberalization and Trade Barriers

Horticultural exports of developing countries are constrained by tariff and non-tariff barriers (NTBs). It is generally agreed that the fruit sector must be prepared to confront a more liberalized trade environment. One of the most significant trade liberalization development and the new trading opportunities and threats affecting horticultural import/export industry worldwide is the recent inclusion of China and Taiwan in the World Trade Organization (WTO).

Tariffs vary by product, season, and country of origin. The restrictive effect of NTBs on horticultural imports is considerably greater than that of tariffs.

A wide range of NTBs include marketing orders such as product specifications, quotas and voluntary export constraints, import licensing, variable levies, minimum price systems, countervail taxes and duties, technical specifications (especially health restrictions, and strict labelling and packaging specifications), and even bureaucratic delays or uncertainties.

Without a clear purpose or scientific justification, there is a tendency to increase the use of NTBs as a discrimination against fresh produce imports. Most of these requirements are voluntary or as recommendation at this stage, but some are soon expected to become mandatory. Among NTBs, fresh produce is particularly heavily constrained by various sanitary and phytosanitary regulation mechanisms on the basis of concerns for consumers and most notably, for the protection/security of the importing country's own production.

Sanitary and phytosanitary regulations can be characterized as trade barriers and WTO requires these to be backed by sound and verifiable scientific evidence. It should be noted that the countries with the most rigorous control procedures are precisely the major exporters, the US, EU and Japan. Almost all national standards either already have guidelines, or are in the process of harmonization with guidelines formulated through international consensus. At this transitional stage, as a result of competition, the market is at the mercy of misinformation and mud sling.

To meet these requirements, the developing exporting countries have to set up expensive, high technology laboratories. Most important of all is that exporting countries will need to provide education, training, and a sophisticated level of technological infrastructure and managerial skill to monitor conditions of production/handling processes for export and to ensure transparency, and allow consultation and dispute settlement processes between trading partners. The costly procedures for the operators to comply with regulatory requirements and restriction tends to be in favour of middle-income developing countries and give new exporting countries only very selective and limited access to enter international trade.

**Changing Consumption Trend and Consumer Demand**

The tropical fruit industry needs to retain the number of loyal consumers who have now moved into the middle and upper income brackets. The industry also needs to expand to new markets with a new consumer group with varying income levels in both the developed and the developing countries.

In addition, recent global economic downturn, falling prices, increased competition among producers, and increased market concentration in favour of supermarkets, have all resulted in a swift of market power in a supply chain that was producer driven, to one that is driven by consumer and their proxy retailer -- a change that has a negative impact on the levels of returns to the producers and exporters.

This has created new opportunities for new companies and pressured some rethinking among existing companies to concentrate their efforts on the core business and move toward more vertical supply chain coordination, in order to improve the overall quality dimensions of the product perceived by the market. The bottom line is to capture the hearts and minds of the consumer.

Emerging New Suppliers and Markets: The expansion of the EU to integrate several countries producing fresh products (Greece, Portugal and Spain) has significantly affected the horticultural exports of developing countries because these members are substantial producers and exporters of horticultural products. The questions facing the non-EU producers are the long-run trends in export supply in and out of the EU by these countries and those that are also potential suppliers to US, Japan and other developing countries in relative competitions with other non-EU member suppliers. The enlargement of EU to include these countries has resulted in the adoption of export standards to EU from one of lenient in the past to one of the strictest.

New emerging suppliers and markets in the region are the result of China and Taiwan's entry into the WTO. China has achieved an outstanding annual horticultural growth rate of 20% per year and that has transformed the country from the fourth largest producer at the start of the decade into the world's largest producer of fruit and vegetables. China has significantly increased its production of the most widely sold varieties of fruit at the international level, namely banana (13% growth per year), oranges (7% per year), apples (26% per year), pears (14% per year), and peaches (15% per year).

This suggests that it has adopted a strategy of supplying the world, and as China's productions meet the regulatory requirements of importing countries and the quality/price expectation of the consumers, it will become an exporting giant. Opportunities still exist at present as China experiences increased purchasing power and improved standard of living.

Thailand has gained an impressive market share of fresh fruit and vegetables imported into China in recent years, mainly tropical fruits such as durian, longan and lychee. A very significant percentage of China's fruit imports are via Hong Kong's re-export trade. During the last 5 years (1996-2000), some 250,000 tons of fresh fruits were re-exported annually to China representing about 25% to 29% of fruit imported to Hong Kong.

Thai fresh fruits such as durian, longan and lychee which constitute 80%, 74%, and 58%, respectively of Hong Kong's total imports of these items in 2000, were exported to China. Demand for longans in China shows maturation for the past years with an annual growth rate of 6% as a result of China's own longan production, in comparison with durian with an annual growth rate of 94% for the past 5 years. Taiwan's recent economic downturn resulting in falling price and increasing strict sanitary and phytosanitary measures have seriously affected Thailand's export prospects into that country.

Better Demand-Driven Information Intelligence: There is an increasing trend for large majority fresh produce wherein importers source their products by purchasing them directly from foreign grower/shippers. The interest of modern trade store has resulted in the growing interrelation between domestic and international market standards and a concentration of market power in the hands of the retail stores.

Another trend is the increasing demand of suppliers in the exporting countries in terms of quality, quantity, delivery times, and efficiency leading to falling in price in the importing countries and a reduction in the profit margins of intermediary agents and producers.

As sourcing by chain stores become global, the potential for targeted marketing strategy to replace mass marketing with the aid of information technology based on better demand information and to share the communication will increase. The quality of tropical produce in the modern retail chains remains mixed and differs widely from poor to excellent.

## Increased Share in Trade of Domestic Market

The retail sector represents about 16% of Thailand's Gross Domestic Product (GDP), and modern trade retailers, concentrating in Bangkok and suburban areas, account for more than 54% of total retail sales. Supermarkets and modern retail chains are gaining market share of fresh produce from traditional wet market distribution systems such as wholesalers, street markets and small stores in many countries' domestic market.

These new stores have emerged to challenge the traditional practice and established status quo of the existing central commodity-based wholesalers. Nowadays, almost all supermarkets place their produce department prominently in the front of the store. The large size of supermarkets permits greater produce diversity, better quality, and convenience required by the time-pressed consumers. While fruit is still sold predominantly whole, fresh vegetables for supermarkets are sold in branded small consumer packs.

An increasing attraction is the offering of more imports, more new varieties, branded products, speciality leafy greens, sprouts, the ready-to-eat hot meal to eat at home and herbal/ health foods and other food services including bakery, drinks, and dairy products. Innovative marketing strategy, technology, and R&D will play an increasing important role.

## Future Focus of Post Harvest Technology

The traditional goals of Post Harvest technology are to improve handling system with emphasis on Post Harvest treatments including storage, packaging and transportation to maintain quality and reduce losses of products between the field and the ultimate destination.

The Post Harvest technology has increasingly become vital in the supply chain management. The medium and long-term challenges are that Post Harvest technology should be more viable to enhance the value chain and to cope with the changing market environments. Following are the new roles of Post Harvest technology:

- To meet the demand-driven legality, safety, and quality requirements;
- To provide mechanisms for combating non-tariff trade barriers and market access;
- Post Harvest technology and quality management system integration in supply chain management;
- Shift to appropriate technologies to serve the modern trade stores and global sourcing by multinational retail chain;
- Targeted/focused technologies to meet the challenge of emerging new supplies and the requirement of the new markets.

# 2

# TECHNOLOGY IN MODERN AGRICULTURAL PRACTICES

Proper Post Harvest processing and handling is an important part of modern agricultural production. Post Harvest processes include the integrated functions of harvesting, cleaning, grading, cooling, storing, packaging, transporting and marketing. The technology of Post Harvest handling bridges the gap between the producer and the consumer—a gap often of time and distance. Post Harvest handling involves the practical application of engineering principles and knowledge of fruit and vegetable physiology to solve problems.

Utilizing improved Post Harvest practices often results in reduced food losses, improved overall quality and food safety, and higher profits for growers and marketers. It is estimated that 9 to 16 percent of the product is lost due to Post Harvest problems during shipment and handling. Mechanical injury is a major cause of losses.

Many of these injuries cannot be seen at the time that the product is packed and shipped, such as internal bruising in tomatoes. Other sources of loss include over-ripening, senescence, the growth of pathogens and the development of latent mechanical injuries.

Many factors contribute to Post Harvest losses in fresh fruits and vegetables. These include environmental conditions such as heat or drouth, mechanical damage during harvesting and handling, improper Post Harvest sanitation, and poor cooling and environmental control. Efforts to control these factors are often very successful in reducing the incidence of disease.

For example, reducing mechanical damage during grading and packing greatly decreases the likelihood of Post Harvest disease because many disease-causing organisms (pathogens) must enter through wounds. Chemicals have been widely used to reduce the incidence of Post Harvest disease. Although effective, many of these materials have been removed from the market in recent years because of economic, environmental, or health concerns.

Increased interest in the proper Post Harvest handling of fresh fruits and vegetables has prompted the widespread use of flumes, water dump tanks, spray washers, and hydrocoolers. To conserve water and energy, most Post Harvest processes that wet the produce recirculate the water after it has passed over the produce.

This recirculated water picks up dirt, trash, and disease-causing organisms. If steps are not taken to prevent their spread, these organisms can infect all the produce that is subsequently processed. In the past, various fungicides and bactericides have been used (alone or in combination with chlorination) to prevent the transmission of diseases. These materials have often been favoured over chlorination because they provide some residual protection after treatment.

## Chlorination

At present, chlorination is one of the few chemical options available to help manage Post Harvest diseases. When used in connection with other proper Post Harvest handling practices, chlorination is effective and relatively inexpensive. It poses little threat to health or the environment The primary objective was to reduce Post Harvest losses of fruits and vegetables during harvesting, farm handling, storage, marketing, transportation and processing.

## The Impact of Post Harvest Losses

Too much of the world's food harvest is lost to spoilage and infestations on its journey to the consumer. In developing countries, where tropical weather and poorly developed infrastructure contribute to the problem, losses are sometimes of staggering proportions. Losses occur in all operations from harvesting through handling, storage, processing and marketing. They vary according to the influence of factors such as the perishability of the commodity; ambient temperature and relative humidity which determine the natural course of decay; fungal and bacterial decay; damage by pests -insects, rodents and birds; the length of time between harvesting and consumption; and practices of Post Harvest handling, storage and processing.

## Post Harvest Disorders or Losses

Post Harvest disorders or losses in quality have economic impacts vastly greater than the actual losses caused by frequency and intensity of their occurrence.

For example there are direct financial losses incurred by the grower from batches of fruit expressing the disorder. Direct losses Horticultural produce is alive and has to stay alive long after harvest. Like other living material it uses up oxygen and gives out carbon dioxide.

It also means that it has to receive intensive care. For a plant, harvesting is a kind of amputation. In the field it is connected to roots that give it water and leaves which provide it with the food energy it needs to live. Once harvested and separated from its sources of water and nourishment it must inevitably die. The role of Post Harvest handling is to delay that death for as long as possible.

## Effect of Temperature on Post Harvest

The ideal temperature often depends on the geographic origin of the product. Tropical plants have evolved in warmer climates and therefore cannot tolerate low temperatures during storage. Plants from tropical origins must be stored above 12°C. This is in contrast to plants which have evolved in temperate, cooler climates which can be stored at 0°C. Table gives a few examples of products and their recommended storage temperature.

***Recommended Storage Temperature a Selection of Fruits and Vegetables***

| *1 - 4 °C* | *5 - 9 °C* | *> 10 °C* |
|---|---|---|
| Apple | Avocado (temperate origin) | Avocado (sub-tropical) |
| Asparagus | Zucchini | Pawpaw |
| Berry fruits | French Bean | Grape fruit and Lemon |
| Broccoli | Passion fruit | Mangoes |
| Peach and Plum | Egg plant | Banana |
| Cherry | Capsicum | Pineapple |
| Grapes | Cucumber | Sweet potato |
| Lettuce | Mandarin orange | Tomato |
| Mushroom | Potato | Pumpkin |
| Carnation | Protea | Ginger |

The rapid cooling after harvest is so important. If the temperature is lowered and the harvested products are put in refrigerated storage, water and quality loss can be reduced. Fresh produce is alive, living and breathing. The general term for all the processes going on inside a living organism is called metabolism. Temperature has a big effect on the rate of metabolism of the product. When the temperature of the product rises, so too does the rate of metabolism.

One of the main processes of metabolism is respiration which is the process of Temperature management for fresh produce is the key to quality. Lowering the temperature as quickly as possible after harvest will slow the rate of metabolism and therefore extend the product's shelf life.

For some flowers, sugar can be added (always with a biocide) to the vase solution to provide carbohydrate, or food for metabolism. As a result, for these types of produce respiration is not limited and the vase life is extended. At extreme of temperature product are damaged. Some suffer chilling injury, some suffer damage at very high temperatures and all products are damaged if they freeze.

Temperature is the most important influence on the rate of deterioration in the quality of produce. High temperatures accelerate ripening and the speed at which rots develop. A 10°C increase in temperature will cause fruit and vegetables to deteriorate twice as fast, as well as encouraging disease organisms to grow twice as fast as well. This is why it is important to remove field heat from the produce as quickly as possible after harvest. Important as it regulates the quantity of refrigerant entering the evaporator coils. When the liquid moves into the evaporator coils the liquid changes into a gas because it absorbs heat energy from the surrounding air (latent heat of vaporisation). As a result the air in the cool room is cooled.

### Post Harvest Losses

Most losses of fresh produce occur between leaving the farm and reaching the consumer. Losses during this period have been estimated to be about 20% of the total crop. These losses may be caused by complete wastage of the product or by lower prices due to a reduction in quality. The cost of these losses is also important as the value of the product increases several fold from the farm gate to the final consumer, so in dollar terms Post Harvest losses are even more significant. There are generally three main causes of Post Harvest losses:

1. Disease caused by fungi and/or bacteria
2. Physical injuries due to insects, mechanical force, chemicals, heat or freezing
3. Non-disease disorders resulting from storage conditions that upset normal metabolism

when the product is rejected further down the marketing chain. Other factors such as cultivar, weather and crop management also influence the development of disease. Some cultivars are more susceptible to diseases than others and wet weather can promote the development of disease and hamper efforts to control outbreaks.

## Essential Oils for the Control of Post Harvest

These plant extracts are generally assumed to be more acceptable and less hazardous than synthetic compounds. This means that essential oils that are registered food grade materials, could be used as alternative anti-fungal and anti-bacterial treatments for fresh produce. The potential for these types of plant extracts is considerable. It is a resource that has not been fully explored.

Post Harvest Laboratory looking at the use of essential oils for the control of Post Harvest pathogens. Essential oils are made up of many different volatile compounds and the make up of the oil quite often varies between species. It seems that the anti-fungal and anti-microbial effects are the result of many compounds acting synergistically.

These means that the individual components by themselves are not as effective. Quite a lot of preliminary work has been done to demonstrate the potential of essential oils for use against Post Harvest pathogens. One work showed that essential oils from red thyme (Thymus zygis), clove buds (Eugenia caryophyllata) and cinnamon leaf (Cinnamomum zeylanicum) prevented the growth of Botrytis cinerea. Other researchers have shown that the essential oil of Monarda citrodora and Melaleuca alternifolia also exhibit antifungal activity against a wide range of common Post Harvest pathogens.

We have looked at the effect of tea tree oil as a vapour on the growth of Botrytis from grapes Tea tree oil has antibacterial and antifungal properties that have secured it a place in the commercial pharmaceutical market. We wanted to determine if these properties may also be useful for the Post Harvest control of fungi on grapes. Our work showed that concentrations of between 100 and 500 ppm were able to prevent the growth of this fungi when it was grown in the laboratory.

## Role of Ethylene in the Post Harvest Life

Ethylene plays a role in the Post Harvest life of many horticultural crops. The important role of ethylene as a plant growth regulator has only been established over the last 50 years but its effects have been known for centuries. The most commonly know use of ethylene is to trigger ripening is some crops, such as bananas and avocados.

The application of ethylene at a controlled rate means that these products can be presented to the customer as "ready to eat". The concentration of ethylene required for the ripening of different products varies. The concentration applied is within the range of 1 and 100 ppm. The time and temperature of treatment also influences the rate of ripening with fruit being ripened at temperatures between 15 to 21°C and relative humidity of 85 - 90 %. Although controlled ripening is the major Post Harvest use of ethylene it can also be applied pre-harvest to promote Post Harvest benefits. The chemical Ethephon produces ethylene and is applied in the field. Ethephon can promote several benefits such as fruit thinning (apples, cherries), fruit loosening prior to harvest (nuts), colour development (apples), degreening (citrus), flower induction (pineapples).

## Other Factors of Post Harvest Losses

In addition to genetic traits, environmental factors such as soil type, temperature, wind during fruit set, frost, and rainy weather at harvest can have adverse effects on storage life, suitability for shipping, and quality. Cultural practices may have dramatic impacts on Post Harvest quality. Good Agricultural Practices during harvest operations and any subsequent Post Harvest handling, minimal or fresh-cut processing, and distribution to consumers must be developed.

## Common Post Harvest Diseases

*Commodity and Disease—Pathogena*

***Apples***

*Blue mold Penicillium expansum (f)*

*Gray mold Botrytis cinerea (f)*

*Black rot Physalospora obtusa (f)*

*Bitter rot Glomerella cingulata (f)*

***Grapes and Small Fruit***

*Blue mold Penicillium sp. (f)*

*Gray mold Botrytis cinerea (f)*

*Rhizopus rot Rhizopus stolonifer (f)*

***Potatoes***

*Fusarium tuber rot Fusarium spp. (f)*

*Wet rot Pythium sp. (f)*

*Bacterial soft rot Erwinia spp. (b)*

*Slimy soft rot Clostridium spp. (b)*

***Peaches and Plums***

*Brown rot Monilinia fructicola (f)*

*Rhizopus rot Rhizopus stolonifer (f)*

*Gray mold Botrytis cinerea (f)*

*Blue mold Penicillium sp. (f)*

*Alternaria rot Alternaria sp. (f)*

*Gilbertella rot Gilbertella persicaria (f)*

***Sweet Potatoes***

*Bacterial soft rot Erwinia chrysanthemi (b)*

*Black rot Ceratocystis fimbriata (f)*

*Ring rot Pythium spp. (f)*

*Java black rot Diplodia gossypina (f)*

*Fusarium surface rot Fusarium oxysporum (f)*

*Fusarium root and stem rot Fusarium solani (f)*

*Rhizopus soft rot Rhizopus nigricans (f)*

*Charcoal rot Marcrophomina sp. (f)*

***Tomatoes and Peppers***

*Alternaria rot Alternaria alternata (f)*

*Buckeye rot Phytophthora sp. (f)*

*Gray mold Botrytis cinerea (f)*

*Soft rot Rhizopus stolonifer (f)*

*Sour rot Geotrichum candidum (f)*

*Bacterial soft rot Erwinia spp. (b) or*

*Pseudomonas spp. (b)*

*Ripe rot Colletotrichum sp. (b)*

***Vegetables in General***

*Watery soft rot Sclerotinia sp. (f)*

*Cottony leak Pythium butleri (f)*

*Fusarium rot Fusarium sp. (f)*

*Bacterial soft rot Erwinia sp. (b) or*

*Pseudomonas spp. (b)*

- f = fungus
- b = bacterium

**Post Harvest Disorders**

Many types of Post Harvest disorders and infectious diseases affect fresh fruits and vegetables (Table). Disorders are the results of stresses related to excessive heat, cold, or improper mixtures of environmental gases such as oxygen, carbon dioxide, and ethylene. Some disorders may be caused by mechanical damage, but all are abiotic in origin (not caused by disease organisms) and cannot be controlled by chlorination or most other Post Harvest chemicals.

However, abiotic disorders often weaken the natural defences of fresh produce, making it more susceptible to biotic diseases those that are caused by disease organisms. Further, in many cases injuries caused by chilling, bruising, sunburn, senescence, poor nutrition, and other factors can mimic biotic diseases.

The control of biotic Post Harvest diseases depends on understanding the nature of disease organisms, the conditions that promote their occurrence, and the factors that affect their capacity to cause losses. Post Harvest diseases may be caused by either fungi or bacteria, although fungi are more common than bacteria in both fruits and vegetables. Post Harvest diseases caused by bacteria are rare in fruits and berries but somewhat more common in vegetables. Viruses seldom cause Post Harvest diseases, although they, like Post Harvest disorders, may weaken the produce.

**Chilling Injury**

Tropical fruits are susceptible to many Post Harvest diseases. They are sensitive to chilling injury, and cannot be kept at low temperatures. It is very important to harvest tropical fruits at the right stage of maturity.

Treatments with sulphur dioxide, fungicides and antioxidants help to reduce losses of tropical fruits from Post Harvest diseases, as does careful attention to humidity and temperature. Most important is careful handling to minimize injuries. Bruised fruit, or fruit with a damaged skin, is more vulnerable to diseases, spoils more quickly and sells more slowly.

Development of the Post Harvest sector commonly requires inputs such as pesticides, cement, galvanized sheeting, wire netting, adequate packaging for the safe handling and transportation of perishables, sprays, boxes, and simple machines. Practical technical interventions are researched and promoted to prevent losses.

For example, storage bins for grain must be cleaned out completely between seasons and disinfected before refuse; shade must be provided for holding perishables together with appropriate containers for their transportation and marketing. Improved technologies for drying fruits, vegetables and root crops have also been introduced to reduce losses arising from seasonal gluts.

**Biological Control**

Biological control of Post Harvest diseases (BCPD) has emerged as an effective alternative. Because wound-invading necrotrophic pathogens are vulnerable to bio-control, antagonists can be applied directly to the targeted area (fruit wounds), and a single application using existing delivery systems (drenches, line sprayers, on-line dips) can significantly reduce fruit decays. The pioneering bio-control products Bio-save and Aspire were registered by EPA in 1995 for control of Post Harvest rots of pome and citrus fruit, respectively, and are commercially available.

The limitations of these bio-control products can be addressed by enhancing bio-control through manipulation of the environment, using mixtures of beneficial organisms, physiological and genetic enhancement of the bio-control mechanisms, manipulation of formulations, and integration of bio-control with other alternative methods that alone do not provide adequate protection but in combination with bio-control provide additive or synergistic effects.

Another critical factor for reducing food losses is quality standards and incentives for delivery of better quality produce through the introduction of a fair and practical grading system. The operation of such a system often calls for training and extension to improve handling, storage, packing, sorting and grading practices.

**Irradiation**

In both domestic and international agricultural markets, expanding the use of irradiation can help to reduce the need for methyl bromide for the Post Harvest control of insect pests. Currently, irradiation treatments have been approved for a variety of food use applications by the U.S. Food and Drug Administration (FDA).

The United States Department of Agriculture (USDA) Animal and Plant Health Inspection Service (APHIS) Plant Protection and Quarantine Service (PPQ) has outlined policy positions regarding the development and use of irradiation treatments for quarantine pest control, and is actively seeking ways to incorporate additional irradiation uses into their plant protection programme (USDA 1995, USDA 1996a, 1996b, 1996c).

Food irradiation is a process by which products are exposed to ionizing radiation to sterilize or kill insects and microbial pests by damaging their DNA.

The FDA permits three types of ionizing radiation to be used on foods: gamma rays from radioactive cobalt-60 and cesium-137, high energy electrons, and x-rays. Although all three have similar effects, gamma rays are most commonly used in food irradiation because of their ability to deeply penetrate pallet loads of food (Forsythe and Evangelou 1993, Morrison 1989).

Gamma irradiation equipment irradiates packaged or bulk commodities by exposing the product to gamma energy from cobalt-60 in closed chambers, which range in size from single modular pallet irradiators to large research or contract irradiation facilities. Absorbed dose is measured as the quantity of radiation imparted per unit of mass of a specified material. The unit of absorbed dose is the gray (Gy) where 1 gray is equivalent to 1 joule per kilogram (ICGFI 1991, NAPPO 1996).

In addition, by interfering with cell division, irradiation inhibits sprouting in tubers, bulbs, and root vegetables (potatoes, onions) and can delay ripening of some tropical fruits, resulting in an extended shelf life for many foods.

In turn, longer shelf lives will enhance trade opportunities between nations by extending time constraints under which fresh produce must be delivered to more distant geographic markets or by allowing the use of slower and less expensive modes of transportation (Kader 1986, Moy 1991, OTA 1985).

Irradiation is used as a pest control tool in over 40 countries, including the United States, Russia, Great Britian and Brazil (Nordion 1995).

The disinfestation of grain as it enters the Soviet Union at the Black Sea Port of Odessa, estimated at over 500,000 metric tons per year, is one of the largest documented commercial industrial applications (Giddings 1991). In the United States, the FDA approved low-doses irradiation for wheat, wheat flour, and potatoes in the early 1960s.

In 1984 and 1985, the FDA approved irradiation of spices and pork, and in the following year, approved low-dose irradiation (up to 1 kGy) to control insects in foods and extend the shelf life of fresh fruits and vegetables (Kader 1986, Morrison 1989). Irradiation has also been used to sterilize food for U.S. hospital patients and astronauts (Morrison 1992). Further, irradiation disinfestation has been found to be effective for treatment of dried fruits, spices, nuts, cut flowers, lumber, and wood chips (ICGFI 1994, Marcotte 1992, Morrison 1989, OTA 1985).

At doses below 1 kGy, irradiation is an effective treatment against various species of fruit flies, mango seed weevils, naval orange worms, potato tuber moths, coddling moths, and other insect species of significance to quarantine situations (Kader 1986).

For irradiation to be approved as a quarantine treatment in the United States, either as a single treatment, or as part of a combined approach (*e.g.*, systems approach), USDA/APHIS/PPQ will require that the level of efficacy be scientifically demonstrated, and that efficacy be demonstrated under commercial settings (USDA 1996).

Factors influencing the response of fresh fruits and vegetables to irradiation include the type of commodity and cultivar, production area and season, maturity at harvest, initial quality, and Post Harvest handling procedures.

Similarly, environmental conditions during irradiation (temperature and atmospheric composition), and dose rates are also influencing factors (ICGFI 1994, Kader 1986, OTA 1985, Morrison 1992).

***Relative Tolerance of Fresh Fruits and Vegetables to Irradiation below 1 kGy***

| | |
|---|---|
| High | Apple, cherry, date, guava, longan, muskmelon, nectarine, papaya, peach, rumbutan, raspberry, strawberry, tamarillo, tomato |
| Medium | Apricot, banana, cherimoya, fig, grapefruit, kumquat, loquat, lychee, orange, passion fruit, pear, pineapple, plum, tangelo, tangerine |
| Low | Avocado, cucumber, grape, green bean, lemon, lime, olive, pepper, sapodilla, soursop, summer squash, leafy vegetables, broccoli, cauliflower |

***Source:*** Kader 1986.

***Comparison of Estimated Post Harvest Treatment Costs for Selected Crops***

| *Crop* | *Methyl Bromide (cents per pound)* | *Irradiation (cents per pound)* |
|---|---|---|
| Strawberries | 0.88 to 0.94 | 2.5 to 8.1 |
| Papaya | 0.88 to 0.94 | 0.9 to 4.2 |
| Mango | 0.88 to 0.94 | Data not available |

***Sources:*** Forsythe and Evalgelou 1993 and 1994, Morrison 1989.

**Fumigation with Methyl Bromide**

The fumigant methyl bromide is generally preferred for the disinfection of both durable and perishable agricultural commodities. This is because of its low phytotoxicity and its wide-ranging action against a large number of pests. Furthermore, large, bulky shipments can be rapidly and easily treated. However, methyl bromide has now been identified as one of the chemicals that are damaging the world's ozone layer. Efforts are being made to develop alternatives, and treatment systems which recycle methyl bromide.

**VHT (Vapour heat treatment)**

Fruitfly is a major problem in the import and export of tropical fruits and vegetables. VHT is an effective, non-chemical method of treatment, widely used to treat mangoes and other tropical fruits grown for export. VHT treatment maintains produce at a certain temperature for a fixed period of time, using a hot water/steam vapour system. This destroys any larvae of the fruitfly which might be present inside the fruit, as well as the eggs and pupae.

***Packaging:*** The packaging of fruits and vegetables should protect them from injury and water loss, and be convenient for handling and marketing. Packages should also provide information about the product, including the grade, handling instructions, and appropriate storage temperatures when the product is on display. The cost of the packaging is important, including whether the container can be recycled or reused.

***Transport:*** Fresh horticultural products should be cooled after harvest and during transport. It is very important that the cold chain is continuous. Once fruit and vegetables have been cooled, they must stay cool. Trucks for road transport may be refrigerated, or may sometimes just be insulated. It is difficult to control the temperature of air shipments, but produce shipped by air should be covered and pre-cooled.

The transportation and storage of fresh fruit and vegetables is an international operation for which the available technology must be used to ensure that produce reaches the consumer in the best possible condition. The use of controlled atmospheric conditions, as a way of reducing the use of chemical preservatives and pesticides, has great potential for the reduction of Post Harvest losses and the maintenance of nutritive value and organoleptic characteristics.

The proper application of controlled atmosphere storage is likely to have as great an impact as the introduction of refrigeration technology a century earlier, yet its potential is only just becoming appreciated, despite its use for apples for many years.

## Quality Considerations

Several factors are important in determining the quality of fruits and vegetables: the appearance, the flavour, the texture, the nutritional value and the safety. Only the first three can be easily identified by consumers.

Successful Post Harvest handling depends partly on the initial quality of the crop at harvest, including the degree of maturity. It also depends on careful handling to minimize mechanical damage, proper management of the environmental conditions, and good sanitation.

Most often, Post Harvest losses are a symptom rather than the problem. Knowledge of their cause is, therefore, essential for deciding measures to prevent them. Such measures may have to be taken by the small farmer, the private trader, a cooperative, the marketing board or other operator, handlers and transporters, wholesale and retail markets, etc.

Future Post Harvest research priorities

- Genetic manipulation for long life, diseases and environmental-stress resistant cultivars.
- Modelling cultivating conditions for high quality and long life (avoiding root stress by heat or drought).
- Environmental friendly pest control.
- Objective determination of suitable harvesting date.
- Post Harvest treatments (heat, UV, irradiation $CO_2$, chemicals) for storage ability.
- Monitoring refrigerating systems
- Optimum storage conditions as storage for tropical fruits, ornamentals, planting material, fresh pack and lightly processed produce.
- Possibilities of modified atmosphere packaging (MAP), absorption layers, inserts sensors, PP films with micropores
- Prevention of possible pathogenic organisms in MAP products.
- Adaptive control of storage conditions with biological sensors.
- Storage during transport and as quarantine measures.
- Humid forced air pre-cooling.
- Minimum impact and vibration norms for bruising during sorting and packaging.
- Objective non destructive measuring of quality and maturity
- Environmental friendly packaging.
- Consumer and traders quality preferences in each country.
- Cost and return of the investment of Post Harvest techny.

- Cost and return of the investment of Post Harvest technologies.
- Fundamental research on senescence, ripening, respiration, ethylene effect, chilling, fermentation, superficial browning.

Major Research Programmes:

- Development of technology for packaging and storage for reduction in Post Harvest losses of fruits and vegetables.
- Development of technology for value added products from fruits, vegetables and food grains.
- Development of machines/equipment such as groundnut pod grader, evaporative cooler, colour sorter, lac scrapper and fruit grader.
- Design development and evaluation of structures for covered crop cultivation of high value crops for varied agro-climatic zones.
- Survey of agro industries, to evaluate the technological status and associated problems.
- Studies on Post Harvest management of fruits and vegetables.
- Adaptive trials on agro-processing centres.

Research in horticulture focuses in the areas of:

- Post Harvest Physiology of horticultural produce
- Impact of pre-harvest and Post Harvest factors on shelf life and Post Harvest quality
- Physiology of fruit ripening
- Ethylene management during ripening and storage of horticultural crops
- Regulation of fruit colour and aroma
- Controlled atmosphere (CA) storage of fruits
- Modified atmosphere packaging of fruits
- Production Technology of fruit crops

- Chemical control of suckers in olives
- Mechanism of Fruitlet Abscission and Fruit Ripening of Mango- with a Particular Reference to Polyamines.
- Regulation of Fruit colour development in apple.
- Genetic transformation of fruit and other horticultural crops

**Conclusion**

Proper evaluation of Post Harvest technologies includes technical, economic and social components. It involves beneficiary participation throughout. Decisions to adopt new or improved technologies are made by individual farmers but their decisions are often strongly influenced by incentives and credit schemes, access to reliable sources of inputs, extension advice, training opportunities, and market information all of which are generally the responsibility of national governments or their agents.

## EXISTING POST HARVEST TECHNOLOGIES

Post Harvest technologies which greatly influence the level of Post Harvest losses and the quality of produce include grading, packaging, Pre-cooling, storage and transportation. Some products also require one or more of the following treatments: trimming, cleaning, curing, disease or insect control, waxing, and ripening. A general description of those technologies adopted in the United States is available in Kader's (1992) book. More detailed but less up-to-date information is also seen in books by Ryall and Lipton (1979), Ryall and Pentzer (1982), and Pantastico (1975). Storage requirements for a wide variety of products are summarized by Hardenburg et al. (1986).

Special requirements for each major product are often published in monographs, such as that for citrus fruits by Wardowski et al. (1986), and for banana by Stover and Simmonds (1987). Kader (1992) also provides long lists of scientific journals, journals with reviews or abstracts and other periodicals which contain up-to-date information of scientific findings and new technologies related to Post Harvest physiology and handling.

This Bulletin is not intended to make a comprehensive review of literature, but will give a brief summary of current technology in common use.

## Grading

Essentially all fruits and vegetables sold in modern markets are graded and sized. Sophisticated marketing systems require precise grading standards for each kind of product. More primitive markets may not use written grade standards, but the products are sorted and sized to some extent. Typical grading facilities in large packinghouses include dumpers and conveyors.

Produce is graded by human eyes and hands, while moving along conveyor belts or rollers. "Electric eyes" are sometimes used to sort produce by colour. In small-scale packing operations, one or a few grading tables may be enough. Dumping, conveying and grasping can cause mechanical injury to some products. Equipment should have a smooth, soft surface, and dumping and grading operations should be gentle, to minimize injuries.

Many products are sized according to their weight. Automated weight sizers of various capacity are used in packinghouses. Round or nearly round fruits are often sized according to their diameter, using automated chain or roller sizers or hand carried ring sizers. An inefficient sizing operation can also cause significant injuries.

## Packaging

There are two very different types of packaging. The first is when produce is packed in containers for transportation and wholesale. The second is when produce is packed into small retail units.

Ideal containers for packing fruits and vegetables should have the following attributes. They are easy to handle, they provide good protection from mechanical damage, they have adequate ventilation, and they are convenient for merchandising (*i.e.* they can easily carry printed information and advertising about the product etc.). They should also be inexpensive, and easily degradable or recyclable. Many kinds of containers have been used, but the "ideal" is yet to be found. Users often put economic considerations first in selecting containers. Fancy containers such as fibreboard boxes, or wooden or plastic crates, are often used for high-value products. Inexpensive containers such as bamboo baskets or nylon net sacs are used for low-priced produce.

Methods of packing can affect the stability of products in the container during shipping, and influence how much the container protects their quality. In fibreboard boxes, for example, delicate and high-priced products are often packed in trays, while other products are simply all put in the box together.

Prepackaging or consumer packaging generally provides additional protection for the products. It is also convenient for retailers as well as customers, and therefore adds value to produce. However, overuse of non-biodegradable plastic trays and wrapping materials, as often seen in modern supermarkets, creates an extra burden of waste disposal and damages the environment.

**Pre-cooling**

Good temperature management is the most effective way to reduce Post Harvest losses and preserve the quality of fruits and vegetables. Desirable storage and transportation temperatures for major fruits and vegetables have been identified and published (*e.g.* Kader 1992, Hardenburg et al. 1986, McGregor 1987). Temperatives which are low, but not low enough to cause chilling-injury, slow down physiological activity and hence the rates of senescence of the products. Low temperatures also reduce microbial growth rates and the rate of decay.

Products harvested from hot fields often carry field heat and have high rates of respiration. Rapid removal of field heat by Pre-cooling is so effective in quality preservation that this procedure is widely used for highly perishable fruits and vegetables. Currently used Pre-cooling methods include room cooling, forced-air cooling, watercooling, vacuum cooling, and package icing. Cooling equipment and technologies for large-scale operations in the United States are well described in Kader's (1992) book. Smaller capacity equipment can be made using the same principles. Liu (1991) has made a brief summary of cooling principles and methods.

Room cooling is a relatively simple method which needs only a refrigerated room with adequate cooling capacity. The produce is packed in containers which are loosely stacked in the cooling room, leaving enough space between containers for each one to be exposed to circulating cold air. The rate of cooling is rather slow compared to other methods of cooling, because the heat inside each container needs to be transferred to the surface of the container by means of conduction before being carried away by the refrigerated air. It may take hours or even days to cool a product, depending on what kind of product it is, the size and nature of the container, and the temperature and velocity of the circulating air.

Forced-air cooling is a more rapid way of using air to cool produce. Cold air is forced to flow through the inside of each container, so that it carries away heat directly from the surface of the produce rather than from the surface of the container. The air flow is produced by creating a pressure difference between the two perforated sides of each container. The containers are stacked inside a covered tunnel with an exhaust fan at one end. Highly perishable and high-value products such as grapes, strawberries and raspberries may be cooled in less than an hour using this method.

Water cooling (also known as hydrocooling), is a rapid and less expensive method. Produce is exposed to cold water by means of showering or dipping. The required cooling time is often a matter of minutes. However, not all kinds of products tolerate hydrocooling. Hydrocooled products inevitably have a wet surface, which may encourage decay in some kinds of produce.

Vacuum cooling is the most efficient way to cool leafy vegetables, particularly headed ones such as head lettuce, cabbage and Chinese cabbage. The produce is placed inside a vacuum tube in which air pressure is reduced. When the pressure is lowered to 4.6 mm Hg, water "boils" off at 0°C from all over the leaf surface. The boiling effect draws heat for vaporization, and hence cools the produce. The cooling time is usually in the order of 20 to 30 minutes. Unfortunately, the equipment needed for vacuum cooling is very expensive, and may not be a good choice for small-scale farming systems.

Package-icing or top-icing is the simplest way of cooling. Adding crushed ice, flake-ice or slurry of ice in containers can cool the produce. However, this method is not suitable for produce which is very sensitive to ice-cold temperatures. Cooling by ice also inevitably wets both the produce and container, and generates water which needs to be drained. Fancy equipment is used in large-scale package-icing operations in the United States (Kader 1992), but many vegetable packers in Taiwan simply add ice flakes to each container, either by hand or by shovel.

**Storage**

Many horticultural crops have a relatively short harvesting season. Storage is needed to extend the marketing period. Various storage methods have been used on a commercial scale (Liu 1991). Air-cooled common storage houses are often used, or underground or cave storage using natural cold air. Storage humidity is sometimes regulated by controlled ventilation and dehumidifiers. Refrigerated storage (cold storage) controls temperature and humidity precisely by mechanical means. Controlled atmosphere (CA) storage controls the concentrations of oxygen and carbon dioxide, in addition to temperature and humidity.

Modified atmosphere (MA) storage also controls oxygen and carbon dioxide concentrations, although not as precisely as in CA, by using semipermeable polymeric films. Ethylene may be scrubbed for products responsive to it, regardless of the storage system used. A good control of temperature, humidity and atmospheric composition maximizes the storage life span of a product.

## AIR-COOLED COMMON STORAGE

This is widely used for storing horticultural products, particularly those which have good keeping quality even without a precise low temperature. However, its use is generally limited to cool seasons in temperate and sub-tropical regions, or high altitude areas where there are low ambient temperatures at night.

An ideal storage room is adequately insulated and has a good ventilation control system which pulls cool air in at night and keeps warm air out during the day. Booth and Shaw (1981) and Hallee and Hunter (undated) have given comprehensive descriptions of how to construct and ventilate air-cooled common storage, but this technology has not been adopted in many parts of the world. For instance, hundreds of thousands of tons of citrus fruits are stored in common storage in Taiwan each winter, but ideal common storage facilities are still at their trial stage. Lack of technical knowledge, and the need for a large investment to build new facilities, delay adoption of this new technology.

Large commercial-scale storage of horticultural products in caves and underground cellars are seen only in northern and central China. The major crops stored are apples, pears and citrus fruits.

Underground or cave storage often provides better temperature control than above-ground air-cooled storage. This method is most useful in areas where there is a long, cold winter, a dry climate, and a thick layer of fine heavy soil. The construction and management technologies developed in China do not yet seem to have spread to other countries. Some modification of Chinese technologies to suit different climatic and soil conditions might well make this technology useful in many parts of the world.

### SA Storage (Controlled Atmosphere)

Commercial application of CA storage is limited to only a few crops, apples and pears being the most popular ones. It is not used for other crops because the benefit is too slight to cover the cost. The technologies involved (Bartsch and Blanpied 1984) are complicated and sophisticated. The cost of building, facilities, and management for CA storage is considerably higher than for refrigerated storage. It should not be recommended for any crop without a thorough cost and benefit analysis.

**MA Storage**

This is a much simpler method than CA. Theoretically, it is possible for MA storage to generate a result similar to that of CA. However, practical experience seems to show that MA storage is usually less reliable and less effective than CA, in terms of extending storage life and preserving quality. Its biggest advantages are that it is easy to use and not expensive. Therefore, commercial trials and application are becoming more common.

Although many research papers have been published on trials using different polymeric films for wrapping or bagging different products, many new experiments are still conducted each year to try and get better results. This method is unlikely to replace CA for crops which are stored in large quantities over long periods, but it is often useful for small-scale and short-period storage of products which benefit from an atmosphere with a low oxygen and high carbon dioxide content.

**Transportation**

Inland transportation of horticultural crops is usually by rail or way truck. Overseas transportation is by sea or by air. A limited amount of high-valued produce is sometimes transported overland by air. The basic requirements for conditions during transportation are similar to those needed for storage, including proper control of temperature and humidity and adequate ventilation (except in the case of MA). In addition, the produce should be immobilized by proper packaging and stacking, to avoid excessive movement or vibration. Vibration and impact during transportation may cause severe bruising or other types of mechanical injury.

CA containers are not used much for transporting fruit and vegetables. Only a small number of CA containers are used for long-distance shipping of highly perishable, high-value products in the United States and for international trade. Refrigerated containers and trailers are more often used for long-distance shipping, whether by sea, rail or truck. Modern technologies for the refrigerated transportation of horticultural produce have been well described by McGregor (1987).

Shipping by refrigerated trucks is not only convenient, but also effective in preserving the quality of product. However, both the initial investment and the operating costs are very high. Another possibility is insulated or properly ventilated trailer trucks. Pre-cooled products can be transported by well insulated non-refrigerated trucks for up to several hours without any significant rise in product temperature. There are considerable cost savings without any sacrifice of quality if trucks are only insulated, rather than refrigerated, for short-distance shipping. If the product is not Pre-cooled or if the shipping distance is long, a ventilated truck is a better choice than an insulated truck without ventilation and without refrigeration. Ventilation alone does not usually provide a uniform cool temperature, but it may help dissipate excessive field heat and respiration heat, and thus avoid high temperature injury.

## DIFFICULTIES IN APPLYING WESTERN TECHNOLOGY

Modern Post Harvest technologies developed over the past twenty or thirty years are now widely used in North America and Europe. Japan is the only Asian country which uses technologies of comparable sophistication.

Although there are many Asian Post Harvest horticulturists who have been trained in the United States, Europe, or Japan and returned to their home countries, technical improvement in many Asian countries has been slow. There are several reasons why Western technologies cannot be applied quickly in Asia. Firstly, Western technologies use sophisticated equipment and facilities which are too large for the small-scale farming systems of Asia. There is some small-scale equipment manufactured in Europe or Japan, but it is too expensive for most Asian countries. Secondly, advantages of applying Western technologies are less obvious in traditional marketing systems than in the supermarkets which dominate retail in western countries.

Large capacity coolers, waxers, grading and packing line facilities, etc. manufactured in the United States are designed for use in large packinghouses. Asian farmers not only have small-scale operations, but also grow diverse crops. Their products are either packed by the growers themselves, or in small packingsheds operated by farmers' cooperatives. The volume each grower or cooperative packs each day is too small for large equipment. Although smaller equipment is often made in European countries and Japan, it is always sold at high prices.

The manufacturers have good reason to ask high prices, because the number of units sold is so small. Reputable manufacturers of farm machinery are reluctant to compromise the quality of their products for the sake of lower prices, so they make only sophisticated, expensive equipment. Since user countries cannot even afford to buy the first unit for trial, it is not possible for them to order many units at once for a bargain price.

Modern supermarkets have refrigerated display cabinets and temporary cold storage. They can keep produce cool continuously, provided it has been properly cooled before arrival. In contrast, traditional markets do not usually have refrigeration. They make every effort to sell produce quickly to reduce display time and avoid quality loss. Even if the produce has been cooled before and during shipping, the "cold chain" breaks at the retail markets and hence reduces the benefit of cooling. Even worse, many consumers mistakenly believe that cold produce has been stored, while warm produce means it is "fresh". This misconception further discourages cooling, which is the most effective way to prevent rapid deterioration of the produce.

## MODIFICATION OF HIGH-TECHNOLOGY

If the capacity of a packing shed is small, only small-scale equipment is needed. If a country cannot afford to buy expensive equipment, cheaper equipment should be developed. In order to lower the price, it may be necessary to sacrifice quality in terms of appearance, convenience, mobility, durability and/or user friendliness. Sacrificing one or more of these attributes might be acceptable at the initial stage of development, provided that the equipment still functions well. In order to encourage users to purchase, the equipment must be sold at an affordable price. Therefore, sophisticated equipment and facilities developed in Western countries may need substantial modification to suit local needs in Asian countries.

This is best done in the user countries. The equipment and facilities needed for Post Harvest handling of horticultural crops are relatively simple. The principles involved have been widely published, and are readily available to farm machinery specialists, while few or no patents are involved. The fundamental need for developing modified equipment of this kind is good cooperation among Post Harvest horticulturists, farm machinery specialists and end users, plus sufficient support from government policies. Several Asian countries have made significant progress in this area in recent years. For example, locally made small-scale forced-air coolers, crushed-ice makers, and small refrigerated storage facilities are often seen at cooperative shipping points in Taiwan today.

## INTEGRATION: THE KEY TO SUCCESS

A typical route for a horticultural product to follow from harvest to arrival in the consumer's hands includes grading, packaging, transportation, wholesaling and retailing. Additional processes which might be added en route are trimming, curing, Pre-cooling, storage, waxing, disease and insect control treatments, and prepackaging. Improper procedures at any step en route may ruin all the efforts made during other steps to preserve the quality of the product. In other words, the final quality of the product is often determined by the worst procedure, which acts as a limiting factor.

Since no Post Harvest technology can remedy quality which is already damaged, a mistake at one point may not be compensated by extra care in other steps. Therefore, an integrated programme covering all Post Harvest procedures is necessary to guarantee success. For instance, if three steps in the Post Harvest handling system need to be improved, they all have to be improved simultaneously. Improving only one or two steps may not be any improvement at all. Taiwan had at least two bad experiences in the past, when it made fragmentary effort to improve Post Harvest technology. One was to import two hydrocoolers from the United States in the 1970s for cooling vegetables produced in central Taiwan. The government subsidized more than half the cost of the equipment, but the users (Farmers' Associations) still could not justify the operational costs for the very limited benefits from using the coolers.

There were no suitable transportation and storage facilities to protect hydrocooled products at that time. The products rewarmed rapidly. Pre-cooled and then rewarmed wet vegetables often decay faster than those which have not been cooled at all. The ultimate fate of the hycrocoolers was to be dismantled and sold as waste metal.

The second failure was the subsidizing of small refrigerated trucks to transport fresh vegetables from eastern Taiwan to Taipei in the 1980s. Since no Pre-cooling facilities were available at the shipping point, the warm vegetables remained warm or became warmer inside the "refrigerated" truck, because the cooling capacity of the truck was enough to keep cool vegetables from warming but not enough to precool warm products.

After these costly lessons of failure, Post Harvest workers realized that improvement programmes should be integrated rather than fragmentary. The people who formulate and supervise a programme must have a comprehensive knowledge of the subject. They should understand all procedures involved, and be able to identify all problems and choose remedial strategies.

The first successful programme in Taiwan began with vegetables in the suburbs. Harvested vegetables were trimmed, hydrocooled, prepackaged, stored in refrigerated rooms until needed, and then shipped to supermarkets in Taipei by insulated small trucks, with or without refrigeration. The hydrocoolers, refrigerated storage, and shipping trucks used were all small and were locally made.

Encouraged by the favorable response of growers and consumers, the programme was extended to major vegetable production areas in central and southern Taiwan. As the volume of vegetables handled under the project increased and shipping distance lengthened, larger coolers, refrigeration storage rooms and shipping trucks came into use.

In order to avoid wetting vegetables, forced-air cooling gradually replaced hydrocooling for some kinds of vegetables and fruit. More and more shipping points adopted the new technology, and Pre-cooled products were not only shipped to supermarkets, but also to large institutional users, and sometimes wholesale markets. Post Harvest losses were reduced, and product value increased.

In order to provide shippers, extension staff and marketing personnel with the essential technical knowledge, a number of short training courses were offered. At the beginning of the programme, Government agencies begged users to try the new technology. Nowadays, it is the users who beg government agencies for newer technology and more technical advice.

## RESEARCH INTO NEW TECHNOLOGIES

Existing technologies for cooling, storage, and transportation of horticultural crops are generally adequate at present. New research should aim at improving equipment, facilities and methods, to make them more efficient and less costly. Applied research for solving specific problems is likely to be more fruitful than basic research in developing countries.

Different crops often need different Post Harvest treatments. Post Harvest horticulturists must establish the requirements for each major crop. Even if optimum conditions and best treatments are known, it is often worthwhile to study the crop's response to less than optimum conditions, since the best treatment is not necessarily the most economic one. For instance, a study was made of optimum storage conditions for Taiwan citrus fruits.

This was followed by experiments in refrigerated storage, improved common storage, and "traditional" common storage. Cave storage is also going to be studied. When these experiments are completed, we shall be able to make recommendations for growers, depending on their specific location and the duration of the storage.

Since improved handling methods often rely heavily on improvements in equipment and facilities, close cooperation between Post Harvest horticulturists and agricultural engineers is necessary.

An area which is urgently in need of new technology is Post Harvest disease and insect control. Effective fungicides for Post Harvest application have become fewer and fewer, since many previously used fungicides are now banned for fear of health hazards. Control of Post Harvest decay must rely on new, safe chemical treatments and practical physical methods, in addition to improved sanitary conditions. Recent findings, such as the use of acetic acid fumigation, (Moyls et al. 1996) and UV-C light (Wilson et al. 1997) indicate the feasibility of finding new chemical and physical methods.

Many countries wish to export their quality products, but have difficulty in fulfilling the quarantine requirements imposed by importing countries. Current quarantine procedures are either chemical or physical (Paull and Armstrong 1994). Many tropical products cannot tolerate any form of cold treatment. Heat treatment is expensive, and often damages the quality of products. Irradiation is a possibility (Lalaguna 1998 Miller and McDonald 1998), but still needs further development. New and safe chemical treatments are yet to be found.

## COORDINATION AND COOPERATION

### More about Taiwan's Experience

Although our past experiences of making improving Post Harvest handling of horticultural crops in Taiwan have not all been successful, significant progress has been made in recent years. In the early stages, even a small success could be realized only after many trials and frustrations. Projects which began in the 1970s ended in almost total failure, because every programme tackled only part of the Post Harvest system programmes were also technically unsound, according to current knowledge. More projects, were launched in the 1980s and some small progress made, but many projects still failed. Eventually the stage was reached in the late 1980s when there were enough high-caliber Post Harvest scientists, while both industry and government agencies were aware of the importance of Post Harvest handling. Taiwan was then able to carry out more effective programmes.

Recent progress has been based on good cooperation between Post Harvest scientists, government coordinators, extension staff, and decision makers at shipping points and terminal markets. Full coordination and cooperation were made possible after repeated discussions and debates. Once the major funding agencies were convinced that an integrated national programme should replace individual fragmentary small projects, the programme really took off.

This type of integrated programmes quickly generated fruitful results. Each year we were able to see some improvement in the quality of fresh produce at the markets, and growers and shippers began to enjoy the benefits of reduced Post Harvest losses and increased profit margins for higher value-added products.

Several steps were crucial in changing programmes from initial failure to recent success:

- A few high-level Post Harvest scientists were trained to serve as programme leaders, while more than a hundred middle-level technical workers were trained to implement projects.
- Key persons were organized into a strong working group.
- Problems, priorities, and programmes were discussed, to generate a consensus within the group.
- Fragmentary projects were integrated into a national programme.

## POST HARVEST HANDLING OF HORTICULTURE PRODUCE

Horticultural crops not only provide us with nutritional and healthy foods, but also generate a cash income to growers. Appropriate production practices, careful harvesting and proper packaging, storage and transport all contribute to the good produce quality. Once a crop is harvested it is impossible to improve its quality. The horticultural crops, because of their high moisture content are inherently more liable to deteriorate especially under tropical conditions. Moreover, they are biologically active and carry out transpiration, respiration, ripening and other biochemical activities, which deteriorate the quality of the produce.

Losses during Post Harvest operations due to improper storage and handling are enormous and can range from 10-40 percent. Post Harvest losses can occur in the field, in packing areas, in storage, during transportation and in the wholesale and retail market. Severe losses occur because of poor facilities, lack of know-how, poor management, market dysfunction or simply the carelessness of farmers. Proper storage conditions, temperature and humidity are needed to lengthen the storage life and maintain quality once the crop has been cooled to the optimum storage temperature.

## CAUSES OF POST HARVEST LOSSES

### Physiological and Biochemical Aspects

The quality of the harvested fruits and vegetables depend on the condition of growth as well as physiological and biochemical changes they undergo after harvest. Fruits and vegetable cells are still alive after harvest and continue their physiological activity. The Post Harvest quality and storage life of fruits appear to be controlled by the maturity. If the fruits are harvested at a proper stage of maturity the quality of the fruits is excellent. Poor quality and uneven ripening are due to early harvesting and late harvesting which results in extremely poor shelf life.

Respiration plays a very significant role in the Post Harvest life of the fruits. In most of the fruits, the rate of respiration increases rapidly with ripening. The sudden upsurge in respiration is called the 'climacteric rise', which is considered to be the turning point in the life of the fruit. After this the senescence and deterioration of the fruit begin. The fruits such as banana, papaya, mango, guava, jackfruit, fig, sapota, etc. belong to the category of climacteric fruits. While litchi, pineapple, grapes, pomegranate, lemon, orange, lime, etc. belong to the non-climacteric group. To extend the Post Harvest life of the fruits its respiration rate should be reduced as far as possible. Thus an understanding of the factors, which influence the rate of respiration, is indispensable to Post Harvest technologies for manipulating the storage behaviour of fruits.

### Mechanical Injury

Owing to their tender texture and high moisture content, fresh fruits and vegetables are very susceptible to mechanical injury. Poor handling, unsuitable containers, improper packaging and transportation can easily cause bruising, cutting, breaking, impact wounding and other forms of injury.

### Parasitic Diseases

High Post Harvest losses are caused by the invasion of fungi, bacteria, insects and other organisms. Microorganisms attack fresh produce easily and spread quickly, because the produce does not have much of a natural defence mechanism and has plenty of nutrients and moisture to support microbial growth. Post Harvest decay control is becoming a more difficult task, because the number of pesticides available is falling rapidly as consumer concern for food safety increases.

## POST HARVEST TECHNOLOGIES

***Pre-cooling:*** Good temperature management is the most effective way to reduce Post Harvest losses and preserve the quality of fruits and vegetables. Products harvested from hot fields often carry field heat and have high rates of respiration. Rapid removal of field heat by Pre-cooling is so effective in quality preservation that this procedure is widely used for highly perishable fruits and vegetables. Currently used Pre-cooling methods include room cooling, forced-air cooling, water cooling, vacuum cooling and package icing

Room cooling is a relatively simple method, which needs only a refrigerated room with adequate cooling capacity. The produce is packed in containers, which are loosely stacked in the cooling room, leaving enough space between containers for each one to be exposed to circulating cold air.

The rate of cooling is rather slow compared to other methods of cooling, because the heat inside each container needs to be transferred to the surface of the container by means of conduction before being carried away by the refrigerated air. It may take hours or even days to cool a product, depending on what kind of product it is, the size and nature of the container, and the temperature and velocity of the circulating air.

Forced-air cooling is a more rapid way of using air to cool produce. Cold air is forced to flow through the inside of each container, so that it carries away heat directly from the surface of the produce rather than from the surface of the container. The airflow is produced by creating a pressure difference between the two perforated sides of each container. The containers are stacked inside a covered tunnel with an exhaust fan at one end. Highly perishable and high-value products such as grapes, strawberries and raspberries may be cooled in less than an hour using this method.

Hydro cooling is a rapid and less expensive method. Produce is exposed to cold water by means of showering or dipping. The required cooling time is often a matter of minutes. However, not all kinds of products tolerate hydro cooling. Hydro cooled products inevitably have a wet surface, which may encourage decay in some kinds of produce.

Vacuum cooling is the most efficient way to cool leafy vegetables, particularly headed ones such as head lettuce, cabbage and Chinese cabbage. The produce is placed inside a vacuum tube in which air pressure is reduced. When the pressure is lowered to 4.6 mm Hg, water "boils" off at 0°C from all over the leaf surface. The boiling effect draws heat for vaporization, and hence cools the produce. The cooling time is usually in the order of 20 to 30 minutes. Unfortunately, the equipment needed for vacuum cooling is very expensive, and may not be a good choice for small-scale farming systems.

Ice bank cooler is a new development in refrigeration with positive ventilation. In this system ice cool air is passed through the boxes containing horticultural produce. This facilitates quicker cooling and large amount of heat is removed in a relatively shorter period. The store maintains a temperature of 0.5 - 0.8°C and relative humidity of 98 percent.

Package icing or top icing is the simplest way of cooling. Adding crushed ice, flake-ice or slurry of ice in containers can cool the produce. However, this method is not suitable for produce, which is very sensitive to ice-cold temperatures. Cooling by ice also inevitably wets both the produce and container, and generates water, which needs to be drained.

## Sanitation

Sanitation is of great concern to produce handlers, not only to protect produce against Post Harvest diseases, but also to protect consumers from food borne illnesses. E.coli 157:$H_7$, Salmonella, Chryptosporidium, Hepatitis and Cyclospera are among the disease causing organisms that have been transferred via fresh fruits and vegetables. Use of a disinfectant in wash water can help to prevent both Post Harvest diseases and food borne illnesses.

Chlorine in the form of a sodium hypochlorite solution or as a dry powdered calcium hypochlorite can be used in hydro-cooling or wash water as a disinfectant. For the majority of vegetables, chlorine in wash water should be maintained in the range of 75-150 ppm (parts per million).

The antimicrobial form, hypochlorous acid, is mostly available in water with a neutral pH (6.5 to 7.5). Organic growers must use chlorine with caution, as it is classified as a restricted material.

Ozonation is another technology that can be used to sanitize produce. A naturally occurring molecule, ozone is a powerful disinfectant. Fruit and vegetable growers have begun using it in dump tanks as well, where it can be thousands of times more effective than chlorine. Ozone not only kills whatever food borne pathogens might be present, it also destroys microbes responsible for spoilage. A basic system consists of an ozone generator, a monitor to gauge and adjust the levels of ozone being produced and a device to dissolve the ozone gas into the water.

Hydrogen peroxide can also be used as a disinfectant. Concentrations of 0.5% or less are effective for inhibiting development of Post Harvest decay caused by a number of fungi. Hydrogen peroxide has a low toxicity rating and is generally recognized as having little potential for environmental damage.

**Presizing and Storage**

For many commodities fruits below a certain size are eliminated manually or mechanically by presizing belt. Undersized fruits are diverted for processing. The sorting process eliminates cull, overripe, misshapen and otherwise defective fruit and separates produce by colour, maturity and ripeness classes.

**Waxing**

Food grade waxes are commonly applied to replace some of the natural waxes removed in the washing and cleaning operations to reduce water loss and to improve appearance. It also provides protection against decay organisms. Waxing may be done after grading and fungicides may be added to the wax. Application of wax and Post Harvest fungicides must be indicated on each container where the refrigerated storage facilities are not available Protective skin coating with wax is one of the methods for increasing the storage life of fresh fruits.

**Packaging**

Packaging of fresh fruits and vegetables has a great significance in reducing the wastage. Packaging provides protection from physical damage during storage, transportation and marketing. There are variety of packages, packaging materials and inserts available.

There are two types of packaging. The first is when produce is packed in containers for transportation and wholesale. The second is when produce is packed into small retail units. Ideal containers for packing fruits and vegetables should have the following attributes.

They are easy to handle, they provide good protection from mechanical damage, they have adequate ventilation and they are convenient for merchandising. They should also be inexpensive and easily degradable or recyclable. Many kinds of containers have been used but the "ideal" is yet to be found. Users often put economic considerations first is selecting containers. Fancy containers such as fibreboard boxes or wooden or plastic crates, are often used for high-value products. Inexpensive containers such as bamboo baskets or nylon net sacs are used for low-priced produce. Methods of packaging can affect the stability of products in the container during shipping, and influence how much the container protects their quality. In fibreboard boxes, for example, delicate and high-priced products are often packed in trays, while other products are simply put in the box in groups.

Prepackaging or consumer packaging generally provides additional protection for the products. It is also convenient for retailers as well as customers, and therefore adds value to produce. However, overuse of non-biodegradable plastic trays and wrapping materials, as often seen in modern supermarkets, which creates an extra burden of waste disposal and damages the environment.

## FACTORS AFFECTING STORAGE LIFE

### Relative Humidity

Transpiration rates (water loss from produce) are determine by the moisture content of the air, which is usually expressed as relative humidity. At high relative humidity, produce maintains salable weight, appearance, nutritional quality and flavour, while wilting, softening and juiciness are reduced. Leafy vegetables with high surface-to-volume ratios; injured produce and immature fruits and vegetables have higher transpiration rates. High temperatures, low relative humidity and high air velocity increase transpiration rates.

Relative humidity needs to be monitored and controlled in storage. Control can be achieved by a variety of methods:

1. Operating a humidifier in the storage area.
2. Regulating air movement and ventilation in relation to storage room load.

3. Maintaining refrigeration coil temperature within the storage room.
4. Using moisture barriers in the insulation of the storage room or transport vehicle.
5. Wetting the storage room floor.
6. Using crushed ice to pack produce for shipment.
7. Sprinkling leafy vegetables, cool-season root vegetables and immature fruits and vegetables with water.

### Temperature

Respiration and metabolic rates are directly related to room temperatures within a given range. The higher the rate of respiration, the faster the produce deteriorates. Lower temperatures reduce respiration rates and the ripening and senescence processes, which prolong the storage life of fruits and vegetables. Low temperatures also slow the growth of pathogenic fungi, which cause spoilage of fruits and vegetables in storage.

Producers should give special care and attention to proper storage conditions for produce with high to extremely high respiration rates, as these crops will deteriorate much more quickly.

It is impossible to make a single recommendation for cool storage of all fruits and vegetables. Climate of the area where the crop originated, the plant part, the season of harvest and crop maturity at harvest are important factors in determining the optimum temperature. A general rule for vegetables is that cool-season crops should be stored at cooler temperatures (0 to 1.7°C) and warm-season crops should be stored at warmer temperatures (7 to 13°C).

### Freezing Injury

Temperatures that are too low can be just as damaging as those too high. Freezing will occur in all commodities below 0°C. Whether injury occurs depends on the commodity. Some can be repeatedly frozen and thawed without damage, while others are ruined by one freezing.

Injury from freezing temperatures can appear in plant tissues as loss of rigidity, softening and water soaking. Injury can be reduced if the produce is allowed to warm up slowly to optimum storage temperatures and if it is not handled during the thawing period. Injured produce should be marketed immediately, as freezing shortens its storage life.

### Chilling Injury

Fruits and vegetables that require warmer storage temperatures (4.5 to 13°C) can be damaged if they are subjected to near-freezing temperatures (0°C). Cooler temperatures interfere with normal metabolic processes. Injury symptoms are varied and often do not develop until the produce has been returned to warmer temperatures for several days. Besides physical damage, chilled produce is often more susceptible to disease infection.

### Ethylene

Ethylene, a natural hormone produced by some fruits as they ripen, promotes additional ripening of produce exposed to it. The old age saying that one bad apple spoils the whole bushel is true. The damaged or diseased fruits produce high levels of ethylene and stimulate the other apples to ripen too quickly. As the fruits ripen, they become more susceptible to disease. Ethylene "producers" should not be stored with fruits, vegetables, or flowers that are sensitive to it.

The result could be loss of quality, reduced shelf life and specific symptoms of injury. Ethylene producers include apples, apricots, avocados, ripening bananas, honeydew melons, papayas, peaches, pears, plums and tomatoes.

## STORAGE FACILITIES

Crops that require different storage conditions will need three different storage facilities.

- Cold Storage (temperatures 0 to 2.2°C)
- Cool Storage (temperatures 4.5 to 13°F)
- Warmer storage (temperatures 13 to 15.6°F)

A recording thermometer can be helpful in determining whether storage facilities are maintaining ideal conditions and are not fluctuating. A maximum/minimum thermometer could be substituted. Relative humidity also should be monitored with a hygrometer.

Controlling and monitoring temperature and relative humidity will enable a grower to maintain optimum storage conditions for maximum storage life of the crop and to minimize crop damage from chilling, freezing or high temperature injuries and water loss from the crop.

### Air-Cooled Common Storage

This is widely used for storing horticultural products, particularly those that have good keeping quality even without a precise low temperature. However, its use is generally limited to cool seasons in temperate and sub-tropical regions, or high altitude areas where there are low ambient temperatures at night. An ideal storage room is adequately insulated and has a good ventilation control system, which draws cool air inside during night and keeps warm air out during the day.

### Refrigerated Storage

Refrigerated storage is a well-established technology widely used for storing horticultural crops all over the world. Its application is limited only by cost and benefit considerations. Essentially, all crops can benefit by being stored at a suitable low temperature, which extends the storage life and preserves quality. Many horticultural crops have storage life spans ranging from less than one month to several months when refrigerated. Therefore, refrigerated storage can be used continuously only if different crops with different harvesting seasons can share the facility. There are other important reasons why this method is not used in many tropical and sub-tropical countries, where refrigeration is needed most. The initial investment cost is too high and its energy consumption too large for many countries.

### Hydrobaric Storage

In this system the horticultural produce is kept in a vacuum-tight and refrigerated container and the air is evacuated by vacuum pump to achieve desired low pressure. The low pressure retards ripening by decreasing respiration. However this is more expensive method.

**Transportation**

Inland transportation of horticultural crops is usually by rail or by truck. Overseas transportation is by sea or by air. A limited amount of high-valued produce is sometimes transported overland by air. The basic requirements for conditions during transportation are proper control of temperature and humidity and adequate ventilation. In addition, the produce should be immobilized by proper packaging and stacking, to avoid excessive movement or vibration. Vibration and impact during transportation may cause severe bruising or other types of mechanical injury. Refrigerated containers and trailers are more often used for long distance shipping, whether by sea, rail or truck.

Shipping by refrigerated trucks is not only convenient, but also effective in preserving the quality of product. However, both the initial investment and the operating costs are very high. Another possibility is insulated or properly ventilated trailer trucks. Pre-cooled products can be transported through well-insulated non-refrigerated trucks for up to several hours without any significant rise in product temperature.

There are considerable cost savings without any sacrifice of quality if trucks are only insulated, rather than refrigerated, for short-distance shipping. If the product is not Pre-cooled or if the shipping distance is long, a ventilated truck is a better choice than an insulated truck without ventilation and without refrigeration. Ventilation alone does not usually provide a uniform cool temperature, but it may help dissipate excessive field heat and respiration heat, and thus avoid high temperature injury.

## POST HARVEST TECHNOLOGY OF HORTICULTURE

In developing countries agriculture is the mainstay of the economy. As such, it should be no surprise that agricultural industries and related activities can account for a considerable proportion of their output. Of the various types of activities that can be termed as agriculturally based, fruit and vegetable processing are among the most important.

Both established and planned fruit and vegetable processing projects aim at solving a very clearly identified development problem. This is that due to insufficient demand, weak infrastructure, poor transportation and perishable nature of the crops, the grower sustains substantial losses. During the Post Harvest glut, the loss is considerable and often some of the produce has to be fed to animals or allowed to rot.

Even established fruit and vegetable canning factories or small/medium scale processing centres suffer huge loss due to erratic supplies.

The grower may like to sell his produce in the open market directly to the consumer, or the produce may not be of high enough quality to process even though it might be good enough for the table. This means that processing capacities will be seriously underexploited.

The main objective of fruit and vegetable processing is to supply wholesome, safe, nutritious and acceptable food to consumers throughout the year.

Fruit and vegetable processing projects also aim to replace imported products like squash, yams, tomato sauces, pickles, etc., besides earning foreign exchange by exporting finished or semi-processed products.

The fruit and vegetable processing activities have been set up, or have to be established in developing countries for one or other of the following reasons:

- diversification of the economy, in order to reduce present dependence on one export commodity;
- government industrialisation policy;
- reduction of imports and meeting export demands;
- stimulate agricultural production by obtaining marketable products;
- generate both rural and urban employment;

- reduce fruit and vegetable losses;
- improve farmers' nutrition by allowing them to consume their own processed fruit and vegetables during the off-season;
- generate new sources of income for farmers/artisans;
- develop new value-added products.

*What Fruit and Vegetables can be Processed?*

Practically any fruit and vegetable can be processed, but some important factors which determine whether it is worthwhile are:

a. The demand for a particular fruit or vegetable in the processed form;

b. The quality of the raw material, *i.e.* whether it can withstand processing;

c. Regular supplies of the raw material.

For example, a particular variety of fruit which may be excellent to eat fresh is not necessarily good for processing. Processing requires frequent handling, high temperature and pressure.

Many of the ordinary table varieties of tomatoes, for instance, are not suitable for making paste or other processed products. A particular mango or pineapple may be very tasty eaten fresh, but when it goes to the processing centre it may fail to stand up to the processing requirements due to variations in its quality, size, maturity, variety and so on.

Even when a variety can be processed, it is not suitable unless large and regular supplies are made available. An important processing centre or a factory cannot be planned just to rely on seasonal gluts; although it can take care of the gluts it will not run economically unless regular supplies are guaranteed.

To operate a fruit and vegetable processing centre efficiently it is of utmost importance to pre-organise growth, collection and transport of suitable raw material, either on the nucleus farm basis or using outgrowers.

## Processing Planning

The secret of a well planned fruit and vegetable processing centre is that it must be designed to operate for as many months of the year as possible. This means the facilities, the buildings, the material handling and the equipment itself must be inter-linked and coordinated properly to allow as many products as possible to be handled at the same time, and yet the equipment must be versatile enough to be able to handle many products without major alterations.

A typical processing centre or factory should process four or five types of fruits harvested at different times of the year and two or three vegetables. This processing unit must also be capable of handling dried/dehydrated finished products, juices, pickles, tomato juice, ketchup and paste, jams, jellies and marmalades, semi-processed fruit products.

Advanced planning is necessary to process a large range of products in varied weather and temperature conditions, each requiring a special set of manufacturing and packaging formulae. The end result of the efforts should be a well-managed processing unit with lower initial investment.

A unit which is sensibly laid out and where one requirement co-relates to another, with a sound costing analysis, leads to an integrated operation.

Instead of over-sophisticated machinery, a sensible simple processing unit may be required when planned production is not very large and is geared mainly to meet the demand of the domestic market.

## Location

The basic objective is to choose the location which minimises the average production cost, including transport and handling.

It is an advantage, all other things being equal, to locate a processing unit near the fresh raw material supply. It is a necessity for proper handling of the perishable raw materials, it allows the processing unit to allow the product to reach its best stage of maturation and lessens injury from handling and deterioration from changes during long transportation after harvesting. An adequate supply of good water, availability of manpower, proximity to rail or road transport facilities and adequate markets are other important requirements.

**Processing Systems**

a. *Small-Scale Processing:* This is done by small-scale farmers for personal subsistence or for sale in nearby markets. In this system, processing requires little investment: however, it is time consuming and tedious. Until recently, small-scale processing satisfied the needs of rural and urban populations. However, with the rising rates of population and urbanisation growth and their more diversified food demands, there is need for more processed and diversified types of food.

b. *Intermediate-Scale Processing:* In this scale of processing, a group of small-scale processors pool their resources. This can also be done by individuals. Processing is based on the technology used by small-scale processors with differences in the type and capacity of equipment used. The raw materials are usually grown by the processors themselves or are purchased on contract from other farmers. These operations are usually located on the production site of in order to assure raw materials availability and reduce cost of transport. This system of processing can provide quantities of processed products to urban areas.

c. *Large-Scale Processing:* Processing in this system is highly mechanised and requires a substantial supply of raw materials for economical operation. This system requires a large capital investment and high technical and managerial skills. Because of the high demand for foods in recent years many large-scale factories were established in developing countries. Some succeeded, but the majority failed, especially in West Africa. Most of the failures were related to high labour inputs and relatively high cost, lack of managerial skills, high cost and supply instability of raw materials and changing governmental policies. Perhaps the most important reason for failure was lack of adequate quantity and regularity of raw material supply to factories. Despite the failure of these commercial operations, they should be able to succeed with better planning and management, along with the undertaking of more in-depth feasibility studies.

It can be concluded that all three types of processing systems have a place in developing countries to complement crop production to meet food demand. Historically, however, small and intermediate scale processing proved to be more successful than large-scale processing in developing countries.

## Choice of Processing Technologies for Developing Countries

FAO maintains (in FAO, 1992c), that the basis for choosing a processing technology for developing countries ought to be to combine labour, material resources and capital so that not only the type and quantity of goods and services produced are taken into account, but also the distribution of their benefits and the prospects of overall growth. These should include:

a. Increasing farmer/artisan income by the full utilisation of available indigenous raw material and local manufacturing of part or all processing equipment;

b. Cutting production costs by better utilisation of local natural resources (solar energy) and reducing transport costs;

c. Generating and distributing income by decentralising processing activities and involving different beneficiaries in processing activities (investors, newly employed, farmers and small-scale industry);

d. maximising national output by reducing capital expenditure and royalty payments, more effectively developing balance-of-payments deficits through minimising imports (equipment, packing material, additives), and maximising export-oriented production;

e. Maximising availability of consumer goods by maximisation of high-quality, standard processed produce for internal and export markets, reducing Post Harvest losses, giving added value to indigenous crops and increasing the volume and quality of agricultural output.

Knowledge and control of the means of production, local manufacturing of processing equipment and development of appropriate/new technologies and more suitable raw material for processing must all be better researched.

Decentralisation of activities must be maintained and coordinated. The introduction of more sophisticated processing equipment and packaging material must be subordinated to internal and export marketing references.

Choosing a technology solely to maximise profits can actually work against true development. Choice should also be based on a solid, long-term market opportunity to ensure viability.

The internal market should be given greater consideration, safeguarded and supported.

Training courses, at all levels, in processing and preservation of indigenous crops, must be expanded.

## CHEMICAL COMPOSITION AND NUTRITIONAL ASPECTS

### General Properties

Fruit and vegetables have many similarities with respect to their compositions, methods of cultivation and harvesting, storage properties and processing. In fact, many vegetables may be considered fruit in the true botanical sense. Botanically, fruits are those portions of the plant which house seeds. Therefore such items as tomatoes, cucumbers, eggplant, peppers, and others would be classified as fruits on this basis.

However, the important distinction between fruit and vegetables has come to be made on an usage basis. Those plant items that are generally eaten with the main course of a meal are considered to be vegetables. Those that are commonly eaten as dessert are considered fruits. That is the distinction made by the food processor, certain marketing laws and the consuming public, and this distinction will be followed in this document.

Vegetables are derived from various parts of plants and it is sometimes useful to associate different vegetables with the parts of the plant they represent since this provides clues to some of the characteristics we may expect in these items.

Fruit as a dessert item, is the mature ovaries of plants with their seeds. The edible portion of most fruit is the fleshy part of the pericarp or vessel surrounding the seeds.

Fruit in general is acidic and sugary. They commonly are grouped into several major divisions, depending principally upon botanical structure, chemical composition and climatic requirements.

Berries are fruit which are generally small and quite fragile. Grapes are also physically fragile and grow in clusters. Melons, on the other hand, are large and have a tough outer rind. Drupes (stone fruit) contain single pits and include such items as apricots, cherries, peaches and plums. Pomes contain many pits, and are represented by apples, quince and pears.

Citrus fruit like oranges, grapefruit and lemons are high in citric acid. Tropical and subtropical fruits include bananas, dates, figs, pineapples, mangoes, and others which require warm climates, but exclude the separate group of citrus fruits.

The compositions of representative vegetables and fruits in comparison with a few of the cereal grains are seen in Table.

***Typical Percentage Composition of Foods from Plant Origin Percentage Composition- Edible Portion***

| *Food* | *Carbo-hydrate* | *Protein* | *Fat* | *Ash* | *Water* |
|---|---|---|---|---|---|
| Cereals | | | | | |
| wheat flour, white | 73.9 | 10.5 | 1.9 | 1.7 | 12 |
| rice, milled, white | 78.9 | 6.7 | 0.7 | 0.7 | 13 |
| maize, whole grain | 72.9 | 9.5 | 4.3 | 1.3 | 12 |
| Earth vegetables | | | | | |
| potatoes, white | 18.9 | 2.0 | 0.1 | 1.0 | 78 |
| sweet potatoes | 27.3 | 1.3 | 0.4 | 1.0 | 70 |

| | | | | | |
|---|---|---|---|---|---|
| Vegetables | | | | | |
| carrots | 9.1 | 1.1 | 0.2 | 1.0 | 88.6 |
| radishes | 4.2 | 1.1 | 0.1 | 0.9 | 93.7 |
| asparagus | 4.1 | 2.1 | 0.2 | 0.7 | 92.9 |
| beans, snap, green | 7.6 | 2.4 | 0.2 | 0.7 | 89.1 |
| peas, fresh | 17.0 | 6.7 | 0.4 | 0.9 | 75.0 |
| lettuce | 2.8 | 1.3 | 0.2 | 0.9 | 94.8 |
| Fruit | | | | | |
| banana | 24.0 | 1.3 | 0.4 | 0.8 | 73.5 |
| orange | 11.3 | 0.9 | 0.2 | 0.5 | 87.1 |
| apple | 15.0 | 0.3 | 0.4 | 0.3 | 84.0 |
| strawberries | 8.3 | 0.8 | 0.5 | 0.5 | 89.9 |

***Source:*** *Anon. (196O)*

Compositions of vegetables and fruit not only vary for a given kind in according to botanical variety, cultivation practices, and weather, but change with the degree of maturity prior to harvest, and the condition of ripeness, which is progressive after harvest and is further influenced by storage conditions. Nevertheless, some generalisations can be made.

Most fresh vegetables and fruit are high in water content, low in protein, and low in fat. In these cases water contents will generally be greater than 70% and frequently greater than 85%.

Commonly protein content will not be greater than 3.5% or fat content greater than 0.5 %. Exceptions exist in the case of dates and raisins which are substantially lower in moisture but cannot be considered fresh in the same sense as other fruit. Legumes such as peas and certain beans are higher in protein; a few vegetables such as sweet corn which are slightly higher in fat and avocados which are substantially higher in fat.

Vegetables and fruit are important sources of both digestible and indigestible carbohydrates. The digestible carbohydrates are present largely in the form of sugars and starches while indigestible cellulose provides roughage which is important to normal digestion. Fruit and vegetables are also important sources of minerals and certain vitamins, especially vitamins A and C. The precursors of vitamin A, including beta-carotene and certain other carotenoids, are to be found particularly in the yellow-orange fruit and vegetables and in the green leafy vegetables.

Citrus fruit are excellent sources of vitamin C, as are green leafy vegetables and tomatoes. Potatoes also provide an important source of vitamin C for the diets of many countries. This is not so much due to the level of vitamin C in potatoes which is not especially high but rather to the large quantities of potatoes consumed.

**Chemical Composition**

Water: Vegetal cells contain important quantities of water. Water plays a vital role in the evolution and reproduction cycle and in physiological processes. It has effects on the storage period length and on the consumption of tissue reserve substances.

In vegetal cells, water is present in following forms:

- bound water or dilution water which is present in the cell and forms true solutions with mineral or organic substances;
- colloidal bound water which is present in the membrane, cytoplasm and nucleus and acts as a swelling agent for these colloidal structure substances; it is very difficult to remove during drying/dehydration processes;
- constitution water, directly bound on the chemical component molecules and which is also removed with difficulty.

Vegetables contain generally 90-96% water while for fruit normal water content is between 80 and 90%.

Mineral substances: Mineral substances are present as salts of organic or inorganic acids or as complex organic combinations (chlorophyll, lecithin, etc.); they are in many cases dissolved in cellular juice.

Vegetables are more rich in mineral substances as compared with fruits. The mineral substance content is normally between 0.60 and 1.80% and more than 60 elements are present; the major elements are: K, Na, Ca, Mg, Fe, Mn, Al, P. Cl, S.

Among the vegetables which are especially rich in mineral substances are: spinach, carrots, cabbage and tomatoes. Mineral rich fruit includes: strawberries, cherries, peaches and raspberries. Important quantities of potassium (K) and absence of sodium chloride (NaCl) give a high dietetic value to fruit and to their processed products. Phosphorus is supplied mainly by vegetables.

Vegetables usually contain more calcium than fruit; green beans, cabbage, onions and beans contain more than 0.1% calcium. The calcium/phosphorus or Ca/P ratio is essential for calcium fixation in the human body; this value is considered normal at 0.7 for adults and at 1.0 for children. Some fruit are important for their Ca/P ratio above 1.0: pears, lemons, oranges and some temperate climate mountain fruits and wild berries.

Even if its content in the human body is very low, iron (Fe) has an important role as a constituent of haemoglobin. Main iron sources are apples and spinach.

Salts from fruit have a basic reaction; for this reason fruit consumption facilitates the neutralisation of noxious uric acid reactions and contributes to the acid-basic equilibrium in the blood.

Carbohydrates: Carbohydrates are the main component of fruit and vegetables and represent more than 90% of their dry matter. From an energy point of view carbohydrates represent the most valuable of the food components; daily adult intake should contain about 500 g carbohydrates.

Carbohydrates play a major role in biological systems and in foods. They are produced by the process of photosynthesis in green plants. They may serve as structural components as in the case of cellulose; they may be stored as energy reserves as in the case of starch in plants; they may function as essential components of nucleic acids as in the case of ribose; and as components of vitamins such as ribose and riboflavin.

Carbohydrates can be oxidised to furnish energy, and glucose in the blood is a ready source of energy for the human body. Fermentation of carbohydrates by yeast and other microorganisms can yield carbon dioxide, alcohol, organic acids and other compounds. Some properties of sugars. Sugars such as glucose, fructose, maltose and sucrose all share the following characteristics in varying degrees, related to fruit and vegetable technology:

- they supply energy for nutrition;
- they are readily fermented by micro-organisms;
- in high concentrations they prevent the growth of micro-organisms, so they may be used as a preservative;
- on heating they darken in colour or caramelise;
- some of them combine with proteins to give dark colours known as the browning reaction.

Some properties of starches:

- They provide a reserve energy source in plants and supply energy in nutrition;
- they occur in seeds and tubers as characteristic starch granules.

Some properties of celluloses and hemicelluloses:

- They are abundant in the plant kingdom and act primarily as supporting structures in the plant tissues;
- they are insoluble in cold and hot water;
- they are not digested by man and so do not yield energy for nutrition;
- the fibre in food which produces necessary roughage is largely cellulose.

Some properties of pectins and carbohydrate gums.

- Pectins are common in fruits and vegetables and are gumlike (they are found in and between cell walls) and help hold the plant cells together;
- pectins in colloidal solution contribute to viscosity of the tomato paste;

- pectins in solution form gels when sugar and acid are added; this is the basis of jelly manufacture.

Fats: Generally fruit and vegetables contain very low level of fats, below 0.5%. However, significant quantities are found in nuts (55%), apricot kernel (40%), grapes seeds (16%), apple seeds (20%) and tomato seeds (18%).

**Organic Acids**

Fruit contains natural acids, such as citric acid in oranges and lemons, malic acid of apples, and tartaric acid of grapes. These acids give the fruits tartness and slow down bacterial spoilage.

We deliberately ferment some foods with desirable bacteria to produce acids and this give the food flavour and keeping quality. Examples are fermentation of cabbage to produce lactic acid and yield sauerkraut and fermentation of apple juice to produce first alcohol and then acetic acid to obtain vinegar.

Organic acids influence the colour of foods since many plant pigments are natural pH indicators.

With respect to bacterial spoilage, a most important contribution of organic acids is in lowering a food's pH. Under anaerobic conditions and slightly above a pH of 4.6, Clostridium botulinum can grow and produce lethal toxins. This hazard is absent from foods high in organic acids resulting in a pH of 4.6 and less.

Acidity and sugars are two main elements which determine the taste of fruit. The sugar/acid ratio is very often used in order to give a technological characterisation of fruits and of some vegetables.

**Nitrogen-containing Substances**

These substances are found in plants as different combinations: proteins, amino acids, amides, amines, nitrates, etc. Vegetables contain between 1.0 and 5.5 % while in fruit nitrogen-containing substances are less than 1% in most cases.

Among nitrogen containing substances the most important are proteins; they have a colloidal structure and, by heating, their water solution above 50°C an one-way reaction makes them insoluble. This behaviour has to be taken into account in heat processing of fruits and vegetables.

From a biological point of view vegetal proteins are less valuable then animal ones because in their composition all essential amino-acids are not present.

**Vitamins**

Vitamins are defined as organic materials which must be supplied to the human body in small amounts apart from the essential amino-acids or fatty acids.

Vitamins function as enzyme systems which facilitate the metabolism of proteins, carbohydrates and fats but there is growing evidence that their roles in maintaining health may extend yet further.

The vitamins are conveniently divided into two major groups, those that are fat-soluble and those that are water-soluble. Fat-soluble vitamins are A, D, E and K. Their absorption by the body depends upon the normal absorption of fat from the diet. Water-soluble vitamins include vitamin C and several members of the vitamin B complex.

***Vitamin-A or Retinol***

This vitamin is found as such only in animal materials - meat, milk, eggs and the like. Plants contain no vitamin A but contain its precursor, beta-carotene. Man needs either vitamin A or beta-carotene which he can easily convert to vitamin A. Beta-carotene is found in the orange and yellow vegetables as well as the green leafy vegetables, mainly carrots, squash, sweet potatoes, spinach and kale.

A deficiency of vitamin A leads to night blindness, failure of normal bone and tooth development in the young and diseases of epithelial cells and membrane of the nose, throat and eyes which decrease the body's resistance to infection.

***Vitamin C***

Vitamin C is the anti-scurvy vitamin. Lack of it causes fragile capillary walls, easy bleeding of the gums, loosening of teeth and bone joint diseases. It is necessary for the normal formation of the protein collagen, which is an important constituent of skin and connective tissue. Like vitamin E, vitamin C favours the absorption of iron. Vitamin C, also known as ascorbic acid, is easily destroyed by oxidation especially at high temperatures and is the vitamin most easily lost during processing, storage and cooking.

Excellent sources of vitamin C are citrus fruits, tomatoes, cabbage and green peppers. Potatoes also are a fair source (although the content of vitamin C is relatively low) because we consume large quantities of potatoes.

**Enzymes**

Enzymes are biological catalysts that promote most of the biochemical reactions which occur in vegetable cells.

Some properties of enzymes important in fruit and vegetable technology are the following:

- in living fruit and vegetables enzymes control the reactions associated with ripening;
- after harvest, unless destroyed by heat, chemicals or some other means, enzymes continue the ripening process, in many cases to the point of spoilage - such as soft melons or overripe bananas;
- because enzymes enter into a vast number of biochemical reactions in fruits and vegetable, they may be responsible for changes in flavour, colour, texture and nutritional properties;
- the heating processes in fruit and vegetables manufacturing/processing are designed not only to destroy micro-organisms but also to deactivate enzymes and so improve the fruit and vegetables' storage stability.

Enzymes have an optimal temperature - around +50°C where their activity is at maximum. Heating beyond this optimal temperature deactivates the enzyme. Activity of each enzyme is also characterised by an optimal pH.

In fruit and vegetable storage and processing the most important roles are played by the enzymes classes of hydrolases (lipase, invertase, tannase, chlorophylase, amylase, cellulase) and oxidoreductases (peroxidase, tyrosinase, catalase, ascorbinase, polyphenoloxidase).

## Turgidity and Texture

The range of textures that are encountered in fresh and cooked vegetables and fruit is indeed great, and to a large extent can be explained in terms of changes in specific cellular components. Since plants tissues generally contain more than two-thirds water, the relationships between these components and water further determine textural differences.

***Cell Turgidity:*** Quite apart from other contributing factors, the state of turgidity, determined by osmotic forces, plays a paramount role in the texture of fruit and vegetables. The cell walls of plant tissues have varying degrees of elasticity and are largely permeable to water and ions as well as to small molecules.

The membranes of the living protoplast are semi-permeable, that is they allow passage of water but are selective with respect to transfer of dissolved and suspended materials.

The cell vacuoles contain most of the water in plant cells and sugars, acids, salts, amino acids, some water-soluble pigments and vitamins, and other low molecular weight constituents are dissolved in this water.

In the living plant, water taken up by the roots passes through the cell walls and membranes into the cytoplasm of the protoplasts and into the vacuoles to establish a state of osmotic equilibrium within the cells.

The osmotic pressure within the cell vacuoles and within the protoplasts pushes the protoplasts against the cell walls and causes them to stretch slightly in accordance with their elastic properties. This is the situation in the growing plant and the harvested live fruit or vegetable which is responsible for desired plumpness, succulence, and much of the crispness.

When plant tissues are damaged or killed by storage, freezing, cooking, or other causes, an important major change that results is denaturation of the proteins of cell membranes resulting in the loss of perm-selectivity. Without perm-selectivity the state of osmotic pressure in cell vacuoles and protoplasts cannot exist, and water and dissolved substances are free to diffuse out of the cells and leave the remaining tissue in a soft and wilted condition.

Other Factors Affecting Texture. The existence of a high degree of turgidity in live fruit and vegetables or whether a relative state of flabbiness develops from loss of osmotic pressure as well as final texture depends on several cell constituents.

Cellulose, Hemicellulose, and Lignin. Cell walls in young plants are very thin and are composed largely of cellulose. As the plant ages cell walls tend to thicken and become higher in hemicellulose and in lignin. These materials are fibrous and tough and are not significantly softened by cooking.

Pectic Substances. The complex polymers of sugar acid derivatives include pectin and closely related substances. The cement-like substance found especially in the middle lamella which helps hold plant cells to one another is a water-insoluble pectic substance.

On mild hydrolysis it yields water-soluble pectin which can form gels or viscous colloidal suspensions with sugar and acid. Certain water-soluble pectic substances also react with metal ions, particularly calcium, to form water-insoluble salts such as calcium pectates. The various pectic substances may influence texture of vegetables and fruits in several ways.

When vegetables or fruit are cooked, some of the water-insoluble pectic substance is hydrolysed into water-soluble pectin. This results in a degree of cell separation in the tissues and contributes to tenderness. Since many fruits and vegetables are somewhat acidic and contain sugars the soluble pectin also tends to form colloidal suspensions which will thicken the juice or pulp of these products.

Fruit and vegetables also contain a natural enzyme which can further hydrolyse pectin to the point where the pectin loses much of its gel forming property. This enzyme is known as pectin methyl esterase. Materials such as tomato juice or tomato paste will contain both pectin and pectin methyl esterase. If freshly prepared tomato juice or paste is allowed to stand the original viscosity gradually decreases due to the action of pectin methyl esterase on pectin gel.

This can be prevented if the tomato products are quickly heated to a temperature of about 82°C (180 F°) to deactivate the pectin methyl esterase liberated from broken cells before it has a chance to hydrolyse the pectin. Such a treatment is commonly practiced in the manufacture of tomato juice products. This is known as the "hot-break process" and yields products of high viscosity.

In contrast, where low viscosity products are desired no heat is used and enzyme activity is allowed to proceed. This is "cold-break" process. After sufficient decrease in viscosity is achieved the product can be heat treated, as in canning, to preserve it for long term storage. It is often also desirable to firm the texture of fruit and vegetables, especially when products are normally softened by processing. In this case advantage is taken of the reaction between soluble pectic substances and calcium ions which form calcium pectates. These calcium pectates are water insoluble and when they are produced within the tissues of fruit and vegetables they increase structural rigidity. Thus, it is common commercial practice to add low levels of calcium salts to tomatoes, apples, and other vegetables and fruits prior to canning or freezing.

## Sources of Colour and Colour Changes

In addition to a great range of textures, much of the interest that fruits and vegetables add to our diets is due to their delightful and variable colours. The pigments and colour precursors of fruit and vegetables occur for the most part in the cellular plastic inclusions such as the chloroplasts and other chromoplasts, and to a lesser extent dissolved in fat droplets or water within the cell protoplast and vacuoles.

These pigments are classified into four major groups which include the chlorophylls, carotenoids, anthocyanins, and anthoanthins. Pigments belonging to the latter two groups also are referred to as flavonoids, and include the tannins.

The Chlorophylls. The chlorophylls are contained mainly within the chloroplasts and have a primary role in the photosynthetic production of carbohydrates from carbon dioxide and water. The bright green colour of leaves and other parts of plants is largely due to the oilsoluble chlorophylls, which in nature are bound to protein molecules in highly organised complexes.

When the plant cells are killed by ageing, processing, or cooking, the protein of these complexes is denatured and the chlorophyll may be released. Such chlorophyll is highly unstable and rapidly changes in colour to olive green and brown. This colour change is believed to be due to the conversion of chlorophyll to the compound pheophytin.

Conversion to pheophytin is favoured by acid pH but does not occur readily under alkaline conditions. For this reason peas, beans, spinach, and other green vegetables which tend to lose their bright green colours on heating can be largely protected against such colour changes by the addition of sodium bicarbonate or other alkali to the cooking or canning water.

However, this practice is not looked upon favourably nor used commercially because alkaline pH also has a softening effect on cellulose and vegetable texture and also destroys vitamin C and thiamin at cooking temperatures.

***The Carotenoids:*** Pigments belonging to this group are fat-soluble and range in colour from yellow through orange to red. They often occur along with the chlorophylls in the chloroplasts, but also are present in other chromoplasts and may occur free in fat droplets. Important carotenoids include the orange carotenes of carrot, maize, apricot, peach, citrus fruits, and squash; the red lycopene of tomato, watermelon, and apricot; the yellow-orange xanthophyll of maize, peach, paprika and squash; and the yellow-orange crocetin of the spice saffron. These and other carotenoids seldom occur singly within plant cells.

A major importance of some of the carotenoids is their relationship to vitamin A. A molecule of orange beta-carotene is converted into two molecules of colourless vitamin A in the animal body. Other carotenoids like alpha-carotene, gamma-carotene, and cryptoxanthin also are precursors of vitamin A, but because of minor differences in chemical structure one molecule of each of these yields only one molecule of vitamin A.

In food processing the carotenoids are fairly resistant to heat, changes in pH, and water leaching since they are fat-soluble. However, they are very sensitive to oxidation, which results in both colour loss and destruction of vitamin A activity.

***The Flavonoids:*** Pigments and colour precursors belonging to this class are water-soluble and commonly are present in the juices of fruit and vegetables. The flavonoids include the purple, blue, and red anthocyanins of grapes, berries, plump, eggplant, and cherry; the yellow anthoxanthins of light coloured fruit and vegetables such as apple, onion, potato, and cauliflower, and the colourless catechins and leucoanthocyanins which are food tannins and are found in apples, grapes, tea, and other plant tissues. These colourless tannin compounds are easily converted to brown pigments upon reaction with metal ions.

Properties of the anthocyanins include a shifting of colours with pH. Thus many of the anthocyanins which are violet or blue in alkaline media become red upon addition of acid.

Cooking of beets with vinegar tends to shift the colour from a purplish red to a brighter red, while alkaline water can influence the colour of red fruits and vegetables toward violet and gray-blue.

The anthocyanins also tend toward the violet and blue hues upon reaction with metal ions, which is one reason for lacquering the inside of metal cans when the true colour of anthocyanin-containing fruits and vegetables is to be preserved.

The water-soluble property of anthocyanins also results in easy leaching of these pigments from cut fruit and vegetables during processing and cooking.

The yellow anthoxanthins also are pH sensitive tending toward a deeper yellow in alkaline media. Thus potatoes or apples become somewhat yellow when cooked in water with a pH of 8 or higher, which is common in many areas. Acidification of the water to pH 6 or lower favours a whiter colour. The colourless tannin compounds upon reaction with metal ions form a range of dark coloured complexes which may be red, brown, green, grey, or black. The various shades of these coloured complexes depend upon the particular tannin, the specific metal ion, pH, concentration of the complex, and other factors not yet fully understood.

Water-soluble tannins appear in the juices squeezed from grapes, apples, and other fruits as well as the brews from extraction of tea and coffee. The colour and clarity of tea are influenced by the hardness and pH of the brewing water. Alkaline waters that contain calcium and magnesium favour the formation of dark brown tannin complexes which precipitate when the tea is cooled.

If acid in the form of lemon juice is added to such tea its colour lightens and the precipitate tends to dissolve. Iron from equipment or from pitted tin cans has caused a number of unexpected colours to develop in products containing tannins, such as coffee, cocoa and foods flavoured with these.

The tannins are also important because they have an astringency which influences flavour and contributes body to such beverages as tea, wine, apple cider, etc.

Fruit and vegetables are in a live state after harvest. Continued respiration gives off carbon dioxide, moisture, and heat which influence storage, packaging, and refrigeration requirements. Continued transpiration adds to moisture evolved and further influences packaging requirements.

Further activities of fruit and vegetables, before and after harvest, include changes in carbohydrates, pectins, organic acids, and the effects these have on various quality attributes of the products.

As for changes in carbohydrates, few generalizations can be given with respect to starches and sugars. In some plant products sugars quickly decrease and starch increases in amount soon after harvest. This is the case for ripe sweet corn which can suffer flavour and texture quality losses in a very few hours after harvest.

Unripe fruit, in contrast, is frequently high in starch and low in sugars. Continued ripening after harvest generally results in a decrease in starch and a increase in sugars as in the case of apples and pears. However, this does not necessarily mean that the starch is the source of the newly formed sugars.

Further, the courses of change in starch and sugars are markedly influenced by Post Harvest storage temperatures. Thus potatoes stored below about 10 C° (50 F°) continue to build up high levels of sugars, while the same potatoes stored above 10 C° do not.

This property is used to help the dehydration process in potato storage. Here potatoes should have a low reducing sugar content so as to minimise Maillard browning reactions during drying and subsequent storage of the dried product. In this case potatoes are stored above 10°C prior to being further processed.

After harvest the pectin changes in fruit and vegetables are more predictable. Generally there is decrease in water-insoluble pectic substance and a corresponding increase in water soluble pectin. This contributes to the gradual softening of fruits and vegetables during storage and ripening. Further breakdown of water-soluble pectin by pectin methyl esterase also occurs.

The organic acids of fruit generally decreases during storage and ripening. This occurs in apples and pears and is especially important in the case of oranges. Oranges have a long ripening period on the tree and time of picking is largely determined by degree of acidity and sugar content which have major effects upon juice quality.

It is important to note that the reduction of acid content on ripening influences more than just the tartness of fruit. Since many of the plant pigments are sensitive to acid, fruit colour would be expected to change. Additionally, the viscosity of pectin gel is affected by acid and sugar contents, both of which change with ripening.

## Stability of Nutrients

One of the principal responsibilities of the food scientist and food technologist is to preserve food nutrients through all phases of food acquisition, processing, storage, and preparation. The key is in the specific sensitivities of the various nutrients, the principles of which are illustrated.

### *Specific Sensitivity and Stability of Nutrients*

| *Nutrient* | *Neutral $p^{H7}$* | *Acid $< p^{H7}$* | *Alkaline $> p^{H7}$* | *Air or Oxygen* | *Light* | *Heat* | *Cooking Losses, Range* |
|---|---|---|---|---|---|---|---|
| | | | **Vitamins** | | | | |
| Vitamin A | S | U | S | U | U | U | 0-40 |
| Ascorbic acid(C) | U | S | U | U | U | U | 0-100 |
| Biotin | S | S | S | S | S | U | 0-60 |
| Carotenes (pro A) | S | U | S | U | U | U | 0-30<br>0-5 |
| Choline | S | S | S | U | S | S | 0-10 |
| Cobalamin ($B_{12}$) | S | S | S | U | U | S | 0-40 |
| Vitamin D | S | | U | U | U | U | 0-10 |
| Essential fatty acids | S | S | U | U | U | S | |
| Folic acid | U | U | S | U | U | U | 0-100 |
| Inositol | S | S | S | S | S | U | 0-95 |
| Vitamin K | S | U | U | S | U | S | 0-5 |
| Niacin (PP) | S | S | S | S | S | S | 0-75 |
| Pantbothenic acid | S | U | U | S | S | U | 0-50 |
| p-Amino Benzoic acid | S | S | S | U | S | S | 0-5 |
| Vitamin $B_6$ | S | S | S | S | U | U | 0-40 |
| Riboflavin ($B_2$) | S | S | U | S | U | U | 0-75 |
| Tbiamin ($B_1$) | U | S | U | U | S | U | 0-80 |
| Tocopherols | S | S | S | U | U | U | 0-55 |

***Source:*** Harris and Karmas, 1975

*(U = Unstable; S = Stable)*

This shows the stability of vitamins, essential amino acids, and minerals to acid, air, light, and heat, and gives an indication of possible cooking losses. Vitamin A is highly sensitive to acid, air, light and heat; vitamin C to alkalinity, air, light and heat; vitamin D to alkalinity, air, light and heat; thiamin to alkalinity, air, and heat in alkaline solutions; etc. Cooking losses of some essential nutrients may be in excess of 75%. In modern food processing operations, however, losses are seldom in excess of 25% .

The ultimate nutritive value of a food results from the sum total of losses incurred throughout its history - from farmer to consumer. Nutrient value begins with genetics of the plant and animal. The farmland fertilization programme affects tissue composition of plants, and animals consuming these plants. The weather and degree of maturity at harvest affect tissue composition.

Storage conditions before processing affect vitamins and other nutrients. Washing, trimming, and heat treatments affect nutrient content. Canning, evaporating, drying, and freezing alter nutritional values, and the choices of times and temperatures in these operations frequently must be balanced between good bacterial destruction and minimum nutrient destruction.

Packaging and subsequent storage affect nutrients. One of the most important factors is the final preparation of the food in the home and the restaurant - the steam table can destroy much of what has been preserved through all prior manipulations.

## Structural Features

The structural unit of the edible portion of most fruits and vegetables is the parenchyma cell. While parenchyma cells of different fruit and vegetables differ somewhat in gross size and appearance, all have essentially the same fundamental structure.

Parenchyma cells of plants differ from animal cells in that the actively metabolising protoplast portion of plant cells represents only a small fraction, of the order of five per cent, of the total cell volume. This protoplast is film-like and is pressed against the cell wall by the large water-filled central vacuole.

The protoplast has inner and outer semi-permeable membrane layers; the cytoplasm and its nucleus are held between them. The cytoplasm contains various inclusions, among them starch granules and plastics such as the chloroplasts and other pigment-containing chromoplasts. The cell wall, cellulose in nature, contributes rigidity to the parenchyma cell and limits the outer protoplasmic membrane. It is also the structure against which other parenchyma cells are cemented to form extensive three-dimensional tissue masses.

The layer between cell walls of adjacent parenchyma cells is referred to as the middle lamella, and is composed largely of pectic and polysaccharide cement-like materials. Air spaces also exist, especially at the angles formed where several cells come together. The relationships between these structures and their chemical compositions are further outlined below. The parenchyma cells will vary in size among plants but are quite large when compared to bacterial or yeast cells. The larger parenchyma cells may have volumes many thousand times greater than a typical bacterial cell.

There are additional types of cells other than parenchyma cells that make up the familiar structures of fruit and vegetables. These include various types of conducting cells which are tube-like and distribute water and salts throughout the plant.

Such cells produce fibrous structures toughened by the presence of cellulose and the wood-like substance lignin. Cellulose, lignin, and pectic substances also occur in specialised supporting cells which increase in importance as plants become older.

An important structural feature of all plants, including fruit and vegetables is protective tissue. This can take many forms but usually is made up of specialised parenchyma cells that are pressed compactly together to form a skin, peel or rind.

Surface cells of these protective structures on leaves, stems or fruit secrete waxy cutin and form a water impermeable cuticle. These surface tissues, especially on leaves and young stems will also contain numerous valve-like cellular structures, the stomata, through which moisture and gases can pass.

## DETERIORATION FACTORS AND THEIR CONTROL

### Enzymic Changes

Enzymes which are endogenous to plant tissues can have undesirable or desirable consequences. Examples involving endogenous enzymes include (a) the Post Harvest senescence and spoilage of fruit and vegetables; (b) oxidation of phenolic substances in plant tissues by phenolase (leading to browning); (c) sugar-starch conversion in plant tissues by amylases; (d) Post Harvest demethylation of pectic substances in plant tissues (leading to softening of plant tissues during ripening, and firming of plant tissues during processing). The major factors useful in controlling enzyme activity are: temperature, water activity, pH, chemicals which can inhibit enzyme action, alteration of substrates, alteration of products and pre-processing control.

### Chemical Changes

***Sensory Quality:*** The two major chemical changes which occur during the processing and storage of foods and lead to a deterioration in sensory quality are lipid oxidation and non-enzymatic browning. Chemical reactions are also responsible for changes in the colour and flavour of foods during processing and storage.

1. Lipid oxidation rate and course of reaction is influenced by light, local oxygen concentration, high temperature, the presence of catalysts (generally transition metals such as iron and copper) and water activity. Control of these factors can significantly reduce the extent of lipid oxidation in foods.
2. Non-enzymic browning is one of the major causes of deterioration which occurs during storage of dried and concentrated foods. The non-enzymic browning, or Maillard reaction, can be divided into three stages: (a) early Maillard reactions which are chemically well-defined steps without browning; (b) advanced Maillard reactions which lead to the formation of volatile or soluble substances; and (c) final Maillard reactions leading to insoluble brown polymers.

## Colour Changes

***Chlorophylls:*** Almost any type of food processing or storage causes some deterioration of the chlorophyll pigments. Phenophytinisation (with consequent formation of a dull olive-brown phenophytin) is the major change; this reaction is accelerated by heat and is acid catalysed.

Other reactions are also possible. For example, dehydrated products such as green peas and beans packed in clear glass containers undergo photo-oxidation and loss of desirable colour.

***Anthocyanins:*** These are a group of more than 150 reddish water-soluble pigments that are very widespread in the plant kingdom. The rate of anthocyanin destruction is pH dependent, being greater at higher pH values. Of interest from a packaging point of view is the ability of some anthocyanins to form complexes with metals such as Al, Fe, Cu and Sn.

These complexes generally result in a change in the colour of the pigment (for example, red sour cherries react with tin to form a purple complex) and are therefore undesirable. Since metal packaging materials such as cans could be sources of these metals, they are usually coated with special organic linings to avoid these undesirable reactions.

***Carotenoids:*** The carotenoids are a group of mainly lipid soluble compounds responsible for many of the yellow and red colours of plant and animal products. The main cause of carotenoid degradation in foods is oxidation. The mechanism of oxidation in processed foods is complex and depends on many factors. The pigments may auto-oxidise by reaction with atmospheric oxygen at rates dependent on light, heat and the presence of pro- and antioxidants.

***Flavour Changes:*** In fruit and vegetables, enzymically generated compounds derived from long-chain fatty acids play an extremely important role in the formation of characteristic flavours. In addition, these types of reactions can lead to significant off-flavours. Enzyme-induced oxidative breakdown of unsaturated fatty acids occurs extensively in plant tissues and this yield characteristic aromas associated with some ripening fruits and disrupted tissues.

The permeability of packaging materials is of importance in retaining desirable volatile components within packages, or in permitting undesirable components to permeate through the package from the ambient atmosphere.

**Nutritional Quality**

The four major factors which affect nutrient degradation and can be controlled to varying extents by packaging are light, oxygen concentration, temperature and water activity. However, because of the diverse nature of the various nutrients as well as the chemical heterogeneity within each class of compounds and the complex interactions of the above variables, generalizations about nutrient degradation in foods will inevitably be broad ones.

***Vitamins: Ascorbic*** acid is the most sensitive vitamin in foods, its stability varying markedly as a function of environmental conditions such as pH and the concentration of trace metal ions and oxygen. The nature of the packaging material can significantly affect the stability of ascorbic acid in foods. The effectiveness of the material as a barrier to moisture and oxygen as well as the chemical nature of the surface exposed to the food are important factors.

For example, problems of ascorbic acid instability in aseptically packaged fruit juices have been encountered because of oxygen permeability of the package and the oxygen dependence of the ascorbic acid degradation reaction.

Also, because of the preferential oxidation of metallic tin, citrus juices packaged in cans with a tin contact surface exhibit greater stability of ascorbic acid than those in enamelled cans or glass containers. The aerobic and anaerobic degradation reactions of ascorbic acid in reduced-moisture foods have been shown to be highly sensitive to water activity, the reaction rate increasing in an exponential fashion over the water activity range of 0.1-0.8.

**Physical Changes**

One major undesirable physical change in food powders is the absorption of moisture as a consequence of an inadequate barrier provided by the package; this results in caking. It can occur either as a result of a poor selection of packaging material in the first place, or failure of the package integrity during storage. In general, moisture absorption is associated with increased cohesiveness.

Anti-caking agents are very fine powders of an inert chemical substance that are added to powders with much larger particle size in order to inhibit caking and improve flowability. Studies in onion powders showed that at ambient temperature, caking does not occur at water activities of less than about 0.4.

At higher activities, however, (aw > 0.45) the observed time to caking is inversely proportional to water activity, and at these levels anti-caking agents are completely ineffective. It appears that while they reduce inter-particle attraction and interfere with the continuity of liquid bridges, they are unable to cover moisture sorption sites.

## Biological Changes

***Microbiological:*** Micro-organisms can make both desirable and undesirable changes to the quality of foods depending on whether or not they are introduced as an essential part of the food preservation process or arise unintentionally and subsequently grow to produce food spoilage.

The two major groups of micro-organisms found in foods are bacteria and fungi, the latter consisting of yeasts and moulds. Bacteria are generally the fastest growing, so that in conditions favourable to both, bacteria will usually outgrow fungi.

Foods are frequently classified on the basis of their stability as non-perishable, semi-perishable and perishable. For example, hermetically sealed and heat processed (*e.g.* canned) foods are generally regarded as non-perishable. However, they may become perishable under certain circumstances when an opportunity for recontamination is afforded following processing.

Such an opportunity may arise if the can seams are faulty, or if there is excessive corrosion resulting in internal gas formation and eventual bursting of the can. Spoilage may also take place when the canned food is stored at unusually high temperatures: thermophilic spore-forming bacteria may multiply, causing undesirable changes such as flat sour spoilage.

Low moisture content foods such as dried fruit and vegetables are classified as semi-perishable. Frozen foods, though basically perishable, may be classified as semi-perishable provided that they are properly stored at freezer temperatures.

The majority of foods (*e.g.* meat and fish, milk, eggs and most fresh fruits and vegetables) are classified as perishable unless they have been processed in some way. Often, the only form of processing which such foods receive is to be packaged and kept under controlled temperature conditions. The species of micro-organisms which cause the spoilage of particular foods are influenced by two factors: a) the nature of the foods and b) their surroundings. These factors are referred to as intrinsic and extrinsic parameters.

The intrinsic parameters are an inherent part of the food: pH, aw, nutrient content, antimicrobial constituents and biological structures. The extrinsic parameters of foods are those properties of the storage environment that affect both the foods and their microorganisms. The growth rate of the micro-organisms responsible for spoilage primarily depends on these extrinsic parameters: temperature, relative humidity and gas compositions of the surrounding atmosphere. The protection of packaged food from contamination or attack by micro-organisms depends on the mechanical integrity of the package (*e.g.* the absence of breaks and seal imperfections), and on the resistance of the package to penetration by micro-organisms.

Metal cans which are retorted after filling can leak during cooling, admitting any microorganisms which may be present in the cooling water, even when the double seam is of a high quality. This fact is widely known in the canning industry and is the reason for the mandatory chlorination of cannery cocling water. Extensive studies on a variety of plastic films and metal foils have shown that microorganisms (including mounds, yeasts and bacteria) cannot penetrate these materials in the absence of pinholes. In practice, however, thin sheets of packaging materials such as aluminium and plastic do contain pinholes.

There are several safeguards against the passage of micro-organisms through pinholes in films:

- because of surface tension effects, micro-organisms cannot pass through very small pinholes unless the micro-organisms are suspended in solutions containing wetting agents and the pressure outside the package is greater than that within;
- materials of packaging are generally used in thicknesses such that pinholes are very infrequent and small;
- for applications in which package integrity is essential (such as sterilisation of food in pouches), adequate test methods are available to assure freedom from bacterial recontamination.

**Macrobiological**

***Insect Pests:*** Warm humid environments promote insect growth, although most insects will not breed if the temperature exceeds about 35 C° or falls below 10 C°. Also many insects cannot reproduce satisfactorily unless the moisture content of their food is greater than about 11%. The main categories of foods subject to pest attack are cereal grains and products derived from cereal grains, other seeds used as food (especially legumes), dairy products such as cheese and milk powders, dried fruits, dried and smoked meats and nuts.

As well as their possible health significance, the presence of insects and insect excrete in packaged foods may render products unsaleable, causing considerable economic loss, as well as reduction in nutritional quality, production of off-flavours and acceleration of decay processes due to creation of higher temperatures and moisture levels.

Early stages of infestation are often difficult to detect; however, infestation can generally be spotted not only by the presence of the insects themselves but also by the products of their activities such as webbing, clumped-together food particles and holes in packaging materials.

Unless plastic films are laminated with foil or paper, insects are able to penetrate most of them quite easily, the rate of penetration usually being directly related to film thickness. In general, thicker films are more resistant than thinner films, and oriented films tend to be more effective than cast films.

The looseness of the film has also been reported to be an important factor, loose films being more easily penetrated than tightly fitted films.

Generally, the penetration varies depending on the basic resin from which the film is made, on the combination of materials, on the package structure, and of the species and stage of insects involved. The relative resistance to insect penetration of some flexible packaging materials is as follows:

- excellent resistance: polycarbonate; poly-ethylene-terephthalate;
- good resistance: cellulose acetate; polyamide; polyethylene (0.254 mm); polypropylene (biaxially oriented); poly-vinyl-chloride (unplasticised);
- fair resistance: acrylonitrile; poly-tetra-fluoro-ethylene; polyethylene (0.123 mm);
- poor resistance: regenerated cellulose; corrugated paper board; kraft paper; polyethylene (0.0254-0.100 mm); paper/foil/polyethylene laminate pouch; poly-vinylchloride (plasticised).

Some simple methods for obtaining insect resistance of packaging materials are as following:

- select a film and a film thickness that are inherently resistant to insect penetration;
- use shrink film over-wraps to provide an additional barrier;
- seal carton flaps completely.

***Rodents:*** Rats and mice carry disease-producing organisms on their feet and/or in their intestinal tracts and are known to harbour salmonella of sero types frequently associated with food-borne infections in humans.

In addition to the public health consequences of rodent populations in close proximity to humans, these animals also compete intensively with humans for food. Rats and mice gnaw to reach sources of food and drink and to keep their teeth short.

Their incisor teeth are so strong that rats have been known to gnaw through lead pipes and unhardened concrete, as well as sacks, wood and flexible packaging materials.

Proper sanitation in food processing and storage areas is the most effective weapon in the fight against rodents, since all packaging materials apart from metal and glass containers can be attacked by rats and mice.

***Summary:*** Major causes of food deterioration include the following:

a. Growth and activities of micro-organisms, principally bacteria, yeasts and moulds;

b. Activities of natural food enzymes;

c. Insects, parasites and rodents;

d. temperature, both heat and cold;

e. Moisture and dryness;

f. Air and in particular oxygen;

g. Light;

h. Time.

Extrinsic factors controlling the rate of food Deterioration reactions are mainly:

a. Effect of temperature;

b. Effect of water activity (aw);

c. Effect of gas atmosphere;

d. Effect of light.

## FRUIT AND VEGETABLE PRESERVATION

### Fresh Storage

***Fresh Fruit and Vegetable Storage:*** Once fruit is harvested, any natural resistance to the action of spoiling micro-organisms is lost. Changes in enzymatic systems of the fruit also occur on harvest which may also accelerate the activity of spoilage organisms. Means that are commonly used to prevent spoilage of fruits must include:

- care to prevent cutting or bruising of the fruit during picking or handling;
- refrigeration to minimise growth of micro-organisms and reduce enzyme activity;
- packaging or storage to control respiration rate and ripening;
- use of preservatives to kill micro-organisms on the fruit.

A principal economic loss occurring during transportation and/or storage of produce such as fresh fruit is the degradation which occurs between the field and the ultimate destination due to the effect of respiration. Methods to reduce such degradation are as follows:

- refrigerate the produce to reduce the rate of respiration;
- vacuum cooling;
- reduce the oxygen content of the environment in which the produce is kept to a value not above 5% of the atmosphere but above the value at which anaerobic respiration would begin. When the oxygen concentration is reduced within 60 minutes the deterioration is in practice negligible.

The following is a summary of some recent developments in Post Harvest technology of fresh fruits and vegetables (Source: Thompson, 1989).

***Harvest Maturity:*** This is particularly important with fruit for export. One recent innovation is the measurement of resonant frequency of the fruit which should enable the grading out of over mature and under-mature fruit before they are packed for export.

***Harvest Method:*** Considerable research is continuing on mechanical harvesting of perishable crops with a view to minimising damage. In fruit trees, controlling their height by use of dwarfing rootstocks, pruning and growth regulating chemicals will lead to easier, cheaper more accurate harvesting.

***Handling Systems:*** Field packing of various vegetables for export has been carried out for many years. In the last decade or so this has been applied, in selected cases, to a few tropical fruit types. Where this system can be practiced it has considerable economic advantages in saving the cost of building, labour and equipment and can result in lower levels of damage into crops.

***Pre-cooling:*** Little innovation has occurred in crop pre-cooling over the last decade. However high velocity, high humidity forced air systems have continued to be developed and refined. These are suitable for all types of produce and are relatively simple to build and operate and, while not providing the speed of cooling of a vacuum or hydrocooler, have the flexibility to be used with almost all crops.

***Chemicals:*** There is a very strong health lobby whose objective is to reduce the use of chemicals in agriculture and particularly during the Post Harvest period. Every year sees the prohibition of the use of commonly used Post Harvest chemicals. New ways need to be developed to control Post Harvest diseases, pest and sprouting.

***Coatings:*** Slowing down the metabolism of fruit and vegetables by coating them with a material which affects their gaseous exchange is being tested and used commercially on a number of products.

***Controlled Environment Transport:*** Recent innovations in this technique have produced great progress as a result of the development and miniaturisation of equipment to measure carbon dioxide and oxygen. Several companies now offer containers where the levels of these two gases can be controlled very precisely.

# 3

# PRESERVATION TECHNIQUES

## Water and Water Activity (aw) in Foods

Microorganisms in a healthy growing state may contain in excess of 80% water. They get this water from the food in which they grow. If the water is removed from the food it also will transfer out of the bacterial cell and multiplication will stop. Partial drying will be less effective than total drying, though for some microorganisms partial drying may be quite sufficient to arrest bacterial growth and multiplication.

Bacteria and yeasts generally require more moisture than moulds, and so moulds often will be found growing on semi-dry foods where bacteria and yeasts find conditions unfavourable; example are moulds growing on partially dried fruits. Slight differences in relative humidity in the environment in which the food is kept or in the food package can make great differences in the rate of microorganism multiplication. Since microorganisms can live in one part of a food that may differ in moisture and other physical and chemical conditions from the food just millimetres away, we must be concerned with conditions in the "micro-environment" of the microorganisms. Thus it is common to refer to water conditions in terms of specific activity.

The term "water activity" is related to relative humidity. Relative humidity is defined as the ratio of the partial pressure of water vapour in the air to the vapour pressure of pure water at the same temperature. Relative humidity refers to the atmosphere surrounding a material or solution.

Water activity or aw is a property of solutions and is the ratio of vapour pressure of the solution compared with the vapour pressure of pure water at the same temperature. Under equilibrium conditions water activity equals:

aw = RH / 100

When we speak of moisture requirements of microorganisms we really mean water activity in their immediate environment, whether this be in solution, in a particle of food or at a surface in contact with the atmosphere.

At the usual temperatures permitting microbial growth, most bacteria require a water activity in the range of about 0.90 to 1.00.

Some yeasts and moulds grow slowly at a water activity down to as low as about 0.65.

Qualitatively, water activity is a measure of unbound, free water in a system available to support biological and chemical reactions. Water activity, not absolute water content, is what bacteria, enzymes and chemical reactants encounter and are affected by at the micro-environmental level in food materials.

Two foods with the same water content can have very different aw values depending upon the degree to which water is free or otherwise bound to food constituents. This is a representative water absorption isotherm for a given food at a given temperature. It shows the final moisture content the food will have when it reaches moisture equilibrium with atmospheres of different relative humidities.

Thus, this food, at the temperature for which this absorption isotherm was established, will ultimately attain a moisture content of 20% at 75% RH (relative humidity). If this food was previously dehydrated to below 20% moisture and placed in an atmosphere of 75% RH, it would absorb moisture until it reached 20%. Conversely, if this food was moistened to greater than 20% water and then placed at 75% RH, it would lose moisture until it reached the equilibrium value of 20%.

Under such conditions some foods may reach moisture equilibrium in the very short time of a few hours, others may require days or even weeks. When a food is in moisture equilibrium with its environment, then the aw of the food will be quantitatively equal to the RH divided by 100.

Qualitatively, water activity is a measure of free or available water, to be distinguished from unavailable or bound water. These states of water also bear a relationship to the characteristic sigmoid shapes of water absorption isotherm curves of various foods.

Thus, according to theory, most of the water corresponding to the portion of the curve below its first inflection point (below 5% moisture) is believed to be tightly bound water, often referred to as an adsorbed mono-molecular layer of water. Moisture corresponding to the region above this point and up to the curve's second inflection point (above 20 % moisture) is thought to exist largely as multi-molecular layers of water less tightly held to food constituent surfaces.

Beyond this second inflection point moisture generally is considered to be largely free water condensed in capillaries and interstices within the food. In this latter portion of the sorption isotherm curve small changes in moisture content result in great changes in a food's aw. The illustrated the moisture sorption isotherms for various dried fruits at 25 °C.

## Preservation by Drying/dehydration

The technique of drying is probably the oldest method of food preservation practiced by mankind. The removal of moisture prevents the growth and reproduction of microorganisms causing decay and minimises many of the moisture mediated deterioration reactions.

It brings about substantial reduction in weight and volume minimising packing, storage and transportation costs and enable storability of the product under ambient temperatures, features especially important for developing countries. The sharp rise in energy costs has promoted a dramatic upsurge in interest in drying worldwide over the last decade.

***Heat and Mass Transfer:*** Dehydration involves the application of heat to vaporise water and some means of removing water vapour after its separation from the fruit/

vegetable tissues. Hence it is a combined/simultaneous (heat and mass) transfer operation for which energy must be supplied. A current of air is the most common medium for transferring heat to a drying tissue and convection is mainly involved.

The two important aspects of mass transfer are:

- the transfer of water to the surface of material being dried and
- the removal of water vapour from the surface.

In order to assure products of high quality at a reasonable cost, dehydration must occur fairly rapidly. Four main factors affect the rate and total drying time:

- the properties of the products, especially particle size and geometry;
- the geometrical arrangement of the products in relation to heat transfer medium (drying air);
- the physical properties of drying medium/ environment;
- the characteristics of the drying equipment.

It is generally observed with many products that the initial rate of drying is constant and then decreases, sometimes at two different rates. The drying curve is divided into the constant rate period and the falling rate period.

Surface area. Generally the fruit and vegetables to be dehydrated are cut into small pieces or thin layers to speed heat and mass transfer. Subdivision speeds drying for two reasons:

- large surface areas provide more surface in contact with the heating medium (air) and more surface from which moisture can escape;
- smaller particles or thinner layers reduce the distance heat must travel to the centre of the food and reduce the distance through which moisture in the centre of the food must travel to reach the surface and escape.

***Temperature:*** The greater the temperature difference between the heating medium and the food the greater will be the rate of heat transfer into the food, which provides the driving force for moisture removal. When the heating medium is air, temperature plays a second important role.

As water is driven from the food in the form of water vapour it must be carried away, or else the moisture will create a saturated atmosphere at the food's surface which will slow down the rate of subsequent water removal. The hotter the air the more moisture it will hold before becoming saturated.

Thus, high temperature air in the vicinity of the dehydrating food will take up the moisture being driven from the food to a greater extent than will cooler air. Obviously, a greater volume of air also can take up more moisture than a lesser volume of air.

***Air Velocity:*** Not only will heated air take up more moisture than cool air, but air in motion will be still more effective. Air in motion, that is, high velocity air, in addition to taking up moisture will sweep it away from the drying food's surface, preventing the moisture from creating a saturated atmosphere which would slow down subsequent moisture removal. This is why clothes dry more rapidly on a windy day.

Some other phenomena influence the drying process and a few elements are summarised below.

***Dryness of Air:*** When air is the drying medium of food, the drier the air the more rapid is the rate of drying. Dry air is capable of absorbing and holding moisture. Moist air is closer to saturation and so can absorb and hold less additional moisture than if it were dry. But the dryness of the air also determines how low a moisture content the food product can be dried to.

***Atmospheric Pressure and Vacuum:*** If food is placed in a heated vacuum chamber the moisture can be removed from the food at a lower temperature than without a vacuum. Alternatively, for a given temperature, with or without vacuum, the rate of water removal from the food will be greater in the vacuum. Lower drying temperatures and shorter drying times are especially important in the case of heat-sensitive foods.

***Evaporation and Temperature:*** As water evaporates from a surface it cools the surface. The cooling is largely the result of absorption by the water of the latent heat of phase change from liquid to gas. In doing this the heat is taken from the drying air or the heating surface and from the hot food, and so the food piece or droplet is cooled.

***Time and Temperature:*** Since all important methods of food dehydration employ heat, and food constituents are sensitive to heat, compromises must be made between maximum possible drying rate and maintenance of food quality.

As is the case in the use of heat for pasteurization and sterilisation, with few exceptions drying processes which employ high temperatures for short times do less damage to food than drying processes employing lower temperatures for longer times.

Thus, vegetable pieces dried in a properly designed oven in four hours would retain greater quality than the same products sun dried over two days. Several drying processes will achieve dehydration in a matter of minutes or even less if the food is sufficiently subdivided.

***Drying Techniques:*** Several types of dryers and drying methods, each better suited for a particular situation, are commercially used to remove moisture from a wide variety of food products including fruit and vegetables.

While sun drying of fruit crops is still practiced for certain fruit such as prunes, figs, apricots, grapes and dates, atmospheric dehydration processes are used for apples, prunes, and several vegetables; continuous processes as tunnel, belt trough, fluidised bed and foam-mat drying are mainly used for vegetables.

Spray drying is suitable for fruit juice concentrates and vacuum dehydration processes are useful for low moisture / high sugar fruits like peaches, pears and apricots.

Factors on which the selection of a particular dryer/ drying method depends include:

- form of raw material and its properties;
- desired physical form and characteristics of dried product;
- necessary operating conditions;
- operation costs.

There are three basic types of drying process:

- sun drying and solar drying;
- atmospheric drying including batch (kiln, tower and cabinet dryers) and continuous (tunnel, belt, belt-trough, fluidised bed, explosion puff, foam-mat, spray, drum and microwave);
- sub-atmospheric dehydration (vacuum shelf/belt/drum and freeze dryers).

The scope has been expanded to include use of low temperature, low energy process like osmotic dehydration. As far dryers are concerned, one useful division of dryer types separates them into air convection dryers, drum or roller dryers, and vacuum dryers. Using this breakdown, The applicability of the more common dryer types to liquid and solid type foods.

***Common Dryer Types Used for Liquid and Solid Doods***

| *Dryer type* | *Usual food type* |
|---|---|
| Air convection dryers | |
| Kiln | pieces |
| cabinet, tray or pan | pieces, purees, liquids |
| Tunnel | pieces |
| continuous conveyor belt | purees, liquids |
| belt trough | pieces |
| air lift | small pieces, granules |
| fluidized bed | small pieces, granules |
| Spray | liquid, purees |
| Drum or roller dryers | |
| Atmospheric | purees, liquids |
| Vacuum | purees, liquids |
| Vacuum dryers | |
| vacuum shelf | pieces, purees, liquids |
| vacuum belt | purees, liquids |
| freeze dryers | pieces, liquids |

***Source:*** Potter, 1984

## Fruit and Vegetable Natural Drying—Sun and Solar Drying

Surplus production and specifically grown crops may be preserved by natural drying for use until the next crop can be grown and harvested. Natural dried products can also be transported cheaply for distribution to areas where there are permanent shortages of fruit and vegetables.

The methods of producing sun and solar dried fruit and vegetables described here are simple to carry out and inexpensive. They can be easily employed by grower, farmer, cooperative, etc.

The best time to preserve fruits and vegetables is when there is a surplus of the product and when it is difficult to transport fresh materials to other markets. This is especially true for crops which are very easily damaged in transport and which stay in good condition for a very short time. Preservation extends the storage (shelf) life of perishable foods so that they can be available throughout the year despite their short harvesting season.

Sun and solar drying of fruits and vegetables is a cheap method of preservation because it uses the natural resource/source of heat: sunlight. This method can be used on a commercial scale as well at the village level provided that the climate is hot, relatively dry and free of rainfall during and immediately after the normal harvesting period. The fresh crop should be of good quality and as ripe (mature) as it would need to be if it was going to be used fresh. Poor quality produce cannot be used for natural drying.

Dried fruit and vegetables have certain advantages over those preserved by other methods. They are lighter in weight than their corresponding fresh produce and, at the same time, they do not require refrigerated storage. However, if they are kept at high temperatures and have a high moisture content they will turn brown after relatively short periods of storage.

Different lots at various stages of maturity (ripeness) must NOT be mixed together; this would result in a poor dried product. Some varieties of fruit and vegetables are better for natural drying than other; they must be able to withstand natural drying without their texture becoming tough so that they are not difficult to reconstitute. Some varieties are unsuitable because they have irregular shape and there is a lot of wastage in trimming and cutting such varieties.

Damaged parts which have been attacked by insects, rodents, diseases, etc. and parts which have been discoloured or have a bad appearance or colour, must be removed. Before trimming and cutting, most fruit and vegetables must be washed in clean water. Onions are washed after they have been peeled.

Trimming includes the selection of the parts which are to be dried, cutting off and disposing of all unwanted material. After trimming, the greater part of the fruit and vegetables cut into even slices of about 3 to 7 mm thick or in halves/quarters, etc.

It is very important to have all slices/parts in one drying lot of the same thickness/size; the actual thickness will depend on the kind of material. Uneven slices or different sizes dry at different rates and this result in a poor quality end product. Onions and root crops are sliced with a hand slicer or vegetable cutter; bananas, tomatoes and other vegetables or fruit are sliced with stainless-steel knives.

As a general rule plums, grapes, figs, dates are dried as whole fruits without cutting/slicing.

Some fruit and vegetables, in particular bananas, apples and potatoes, go brown very quickly when left in the air after peeling or slicing; this discoloration is due to an active enzyme called phenoloxidase. To prevent the slices from going brown they must be kept under water until drying can be started. Salt or sulphites in solution give better protection. However, whichever method is used, further processing should follow as soon as possible after cutting or slicing.

Blanching—exposing fruit and vegetable to hot or boiling water—as a pre-treatment before drying has the following advantages:

- it helps clean the material and reduce the amount of microorganisms present on the surface;

- it preserves the natural colour in the dried products; for example, the carotenoid (orange and yellow) pigments dissolve in small intracellular oil drops during blanching and in this way they are protected from oxidative breakdown during drying;
- it shortens the soaking and/or cooking time during reconstitution.

During hot water blanching, some soluble constituents are leached out: water-soluble flavours, vitamins (vitamin C) and sugars. With potatoes this may be an advantage as leaching out of sugars makes the potatoes less prone to turning brown.

Blanching is a delicate processing step; time, temperature and the other conditions must be carefully monitored. A suitable water-blanching method in traditional processing is as follows:

- the sliced material is placed on a square piece of clean cloth; the corners of the cloth are tied together;
- a stick is put through the tied corners of the cloth;
- the cloth is dipped into a pan containing boiling water and the stick rests across the top of the pan thus providing support for the cloth bag.

The average blanching time is 6 minutes. The start of blanching has to be timed from the moment the water starts to boil again after the cloth bag has been dipped into the pan. While the material is being blanched the cloth bag should be raised and lowered in the water so that the material is heated evenly. When the blanching time is completed the cloth bag and its content should be dipped into cold water to prevent over-blanching. If products are over-blanched (boiled for too long) they will stick together on the drying trays and they are likely to have a poor flavour.

Green beans, carrots, okra, turnip and cabbage should always be blanched. The producer can choose whether or not potatoes need blanching. Blanching is not needed for onions, leeks, tomatoes and sweet peppers. Tomatoes are dipped into hot water for one minute when they need to be peeled but this is not blanching.

## Use of Preservatives

Preservatives are used to improve the colour and keeping qualities of the final product for some fruits and vegetables. Preservatives include items such as sulphur dioxide, ascorbic acid, citric acid, salt and sugar and can either be simple or compound solutions. Treatment with preservatives takes place after blanching or, when blanching is not needed, after slicing. In traditional, simple processing the method recommended is:

- put enough preservative solution to cover the cloth bag into a container/pan;
- dip the bag containing the product into the preservative solution for the amount of time specified;
- remove the bag and put it on a clean tray while the liquid drains out. The liquid which drains out must not go back into the preservative solution because it would weaken the solution.

Care must be taken after each dip to refill the container to the original level with fresh preservative solution of correct strength. After five lots of material have been dipped, the remaining solution is thrown away; *i.e.* a fresh lot of preservative solution is needed for every 5 lots of material. The composition and strength of the preservative solution vary for different fruit and vegetables.

The strength of sulphur dioxide is expressed as "parts per million" (ppm). 1.5 grams of sodium metabisulphite in one litre of water gives 1000 ppm of sulphur dioxide. Details for solutions of different strengths are given in the following table.

Sodium bicarbonate is added to the blanching water when okra, green peas and some other green vegetables are blanched. The chemical raises the pH of the blanching water and prevents the fresh green colour of chlorophyll being changed into pheophytin which is unattractive brownish-green.

The preservative solutions in the fruit and vegetable pre-treatment can only be used in enamelled, plastic or stainless-steel containers; never use ordinary metal because solutions will corrode this type of container.

### *Dilutions of Sodium Metabisulphite with Water to Obtain "PM" of Sulphur Dioxide ($SO_2$)*

| *PPM $SO_2$* | *Sodium Metabisulphite* | |
|---|---|---|
| | *Grams per litre of water* | *Grams per 20 litre tin of water* |
| 1000 | 1.5 | 30.0 |
| 2000 | 3.0 | 60.0 |
| 3000 | 4.5 | 90.0 |
| 4000 | 6.0 | 120.0 |
| 5000 | 7.5 | 150.0 |
| 6000 | 9.0 | 180.0 |
| 7000 | 10.5 | 210.0 |

One level teaspoon of sodium metabisulphite = c. 5 g.

As a general rule, preservatives are not used for treating onions, garlic, leeks, chillies and herbs.

**Osmotic Dehydration**

In osmotic dehydration the prepared fresh material is soaked in a heavy (thick liquid sugar solution) and/or a strong salt solution and then the material is sun or solar dried. During osmotic treatment the material loses some of its moisture. The syrup or salt solution has a protective effect on colour, flavour and texture.

This protective effect remains throughout the drying process and makes it possible to produce dried products of high quality. This process makes little use of sulphur dioxide.

**Sun Drying**

The main problems for sun drying are dust, rain and cloudy weather. Therefore, drying areas should be dust-free and whenever there is a threat of a dust storm or rain, the drying trays should be stacked together and placed under cover.

In order to produce dust-free and hygienically clean products, fruit and vegetable material should be dried well above ground level so that they are not contaminated by dust, insects, livestock or people. All materials should be dried on trays designed for the purpose; the most common drying trays have wooden frames with a fitted base of nylon mosquito netting. Mesh made of woven grass can also be used. Metal netting must NOT be used because it discolours the product.

The trays should be placed on a framework at table height from the ground. This allows the air to circulate freely around the drying material and it also keeps the food product well away from dirt. Ideally the area should be exposed to wind and this speed up drying, but this can only be done if the wind is free of dust.

With 80 cm x 50 cm trays, the approximate load for a tray is 3 kg; the material should be spread in even layers. During the first part of the drying period, the material should be stirred and turned over at least once an hour.

This will help the material dry faster and more evenly, prevent it sticking together and improve the quality of the finished product. Products for sun drying should be prepared early in the day; this will ensure that the material enjoys the full effect of the sun during the early stages of drying.

At night the trays should be stacked in a ventilated room or covered with canvas. Plastic sheets should NEVER be used for covering individual trays during sun drying.

Dry or nearly dry products can be blown out of the tray by the wind. However, this can be protected by covering the loaded tray with an empty one; this also gives protection against insects and birds.

## Shade Drying

Shade drying is carried out for products which can lose their colour and/or turn brown if put in direct sunlight. Products which have naturally vivid colours like herbs, green and red sweet peppers, chillies, green beans and okra give a more attractive end-product when they are dried in the shade.

The principles for the shade drying are the same as for sun drying. The material to be dried requires full air circulation. Therefore, shade drying is carried out under a roof or thatch which has open sides; it cannot be done either inside conventional buildings with side walls or in compounds sheltered from wind.

Under dry conditions when there is a good circulation of air, shade drying takes little more time than is normally required for drying in full sunlight.

Identification of suitable designs of solar dryers for different applications:

In the selection of appropriate solar dryers for commercial scale operation, it is imperative that economics be kept in view at all time. A total Energy System concept should be employed and due consideration be given to parasitic energy consumption.

The following features have been identified:

- large scale dryers are more promising than small scale ones. However, small scale dryers should not be neglected.
- the dryer should be designed to maximise the utilisation factor of the capital investment, *i.e.* multi-products (fruit, vegetables and other raw material) and multi-use (*e.g.* drying and heating water for domestic use).
- in general, an auxiliary heat source should be provided to assure reliability, to handle peak loads and also to provide continuous drying during periods of no sunshine.
- forced convection indirect dryers are preferred because they offer better control, more uniform drying and because of their high heat collection efficiency result in smaller collector area. However, parasitic power should be kept to a minimum.

Two dryer systems have been identified:

- a cabinet type dryer with natural convection for internal air circulation for the processing of dried fruit such as mango, banana, pineapple, apricot, pear, apple, etc. and also for potato chips and other vegetables;
- a greenhouse type dryer with forced air circulation.

Some of the barriers to the commercial development of solar dryers have been attributed to:

- initial cost—poor farmers cannot afford them
- durability—constant breakdown due to using low cost building materials
- misuse—through lack of training and technical skills
- dependability and reliability - during the wet season when drying is critical there is not enough solar energy available
- the wider use of Solar Drying Systems has been limited by other factors which are not necessarily of a technical or technological nature. Among the most important are the lack of national policies directed to promoting the drying of produce at the production site, in order to reduce losses, improve quality and increase farmers' earnings.

## Construction of Solar Dryers

In the case of simple natural convection dryers it may be more appropriate to build and operate a number of small units. Multiplicity allows diversity, since more than one crop can be dried at a time. A further advantage is that if one dryer is out of operation due to damage, drying can still continue at reduced capacity using the other dryers.

On the other hand, more sophisticated dryers, such as forced convection solar dryers, benefit from economies of scale due to the investment tied up in the fan and the source of heat.

Generally speaking, one large dryer will be more cost-effective than two smaller units. However, it should be taken into consideration that an oversized unit will be operating at less than full capacity, reducing any cost advantage. The drying area required will depend on local conditions, commodity and number of trays on each rack or trolley.

Construction methods and materials: Construction methods and available materials may vary considerably from location to location. It is not within the scope of this document to discuss individual, local circumstances. Some general guidelines regarding factors which must be considered can, however, be given: dimensions of standard materials. Where possible, design should take account of the sizes of material locally available. For example, it would be poor design to specify the width of a corrugated iron collector as 1.1 m if the standard width of a corrugated iron sheet is 1 m. Before finalising a design the commercial availability of materials must be ascertained.

***Use of Rural Materials:*** The cost of building of solar dryer can be minimised if the producer is able to use wood cut straight from the forest rather than prepared timber.

Careful design in the development stage of a dryer can often facilitate the use of cheaper materials. Difficulties caused by these materials are in joining pieces of the structure, in sealing the structure against air leaks, and in attaching the plastic sheet to the (wooden) frame.

There is obvious scope for designs which use prepared timber for strategic points and unprepared at others.

Where the use of wood is necessary, remember to take environmental factors into consideration. For example, determine the effect of flash flooding or termites might be and take the appropriate preventive action.

***Use of Plastic Sheets:*** For many solar dryers, the clear plastic sheet used is the major capital cost to the farmer; therefore, the type of plastic chosen is important.

A choice must be made between a relatively cheap plastic such as ordinary polyethylene which will last, at best, for one season due to photo-degradation and wear and tear; and a more expensive, better quality plastic less prone to photo-degradation; or even glass or a rigid plastic.

Attaching plastic sheet to the framework structure, so as to minimise the likelihood of the plastic being torn is, perhaps, the most difficult part of building a dryer. Listed below are some general points which should be followed to prolong the useful lifetime of plastic sheet on a solar dryer:

- when attaching plastic sheet to the framework, care should be taken to stretch the plastic at the points of attachment, but the plastic should not be so loose that it will flap about in the wind;

- rather than merely stapling or nailing the plastic directly to the framework, it is preferable to sandwich the plastic between the framework and a batten. This may not be practical when unprepared wood or other materials are being used.
- no sharp edges should come in contact with the plastic sheet since these will initiate tears;
- fold over the plastic at the point of attachment to the frame, so that there are two or more layers of plastic. This will help prevent tears;
- when fixing the sheet over the framework, sags and hollows in which water can collect should be avoided wherever possible;
- the dryer should be handled as carefully and as seldom as possible during operation and when not in use.

***Technical Criteria:*** The following design factors must be established:

- the throughput of the dryer over the productive season; the size of batch to be dried;
- the drying period(s) under stated conditions;
- the initial and desired final moisture content of the commodity (if known);
- the drying characteristics of the commodity, such as maximum drying temperature, effect of sunlight upon the product quality, etc.;
- climatic conditions during the drying season, *i.e.* sunlight intensity and duration; air temperature and humidity; wind speed (such data may be available from local meteorological stations);
- availability and reliability of electrical power;
- the availability, quality, durability and price of potential construction materials such as:
  - o glazing materials: glass, plastic sheet or film;
  - o wood (prepared or unprepared);

- o nails, screw, bolts, etc.;
- o metal sheet, flat or corrugated angle iron;
- o bricks (burnt or mud), concrete blocks, stones, cement, sand, etc.
- o roofing thatch;
- o metal mesh, wire netting, etc.
- o mosquito netting, muslin, etc.
- o bamboo or fibre weave;
- o black paint, other blackening materials;
- o insulation material; sawdust, etc.;

- the type of labour available to build and operate the dryer;
- the availability of clean water at the site for preparation of the commodity prior to drying.

In any one situation there may well be other technical factors that need to be considered.

***Socio-economic Criteria:*** From the initial considerations, estimates of the capital costs of the dryer, the price of the commodity to be dried, and the likely selling price of the dried product will have been made. Other question that need to be considered are the following:

- who will own the dryer?
- is the dryer to be constructed by the end-user (with or without advice from extension agencies), local contractors, or other organisations?
- who will operate and maintain it?
- how can the drying operation be incorporated into current practices?
- are sources of finance from local authorities or extension agencies available, etc.?

Obviously there are many other socio-economic factors, particularly those of a local nature, which must be taken into account. It cannot be stressed too highly that if such factors are not taken into account and evaluated, then is every chance that an inappropriate dryer design may result. Equal emphasis must be placed on both technical and socio-economic factors.

**Summary**

(a) Situations where solar dryer may be useful:

- where the cost of conventional energy is prohibitive and/or the supply is erratic, to supplement existing artificial drying systems and reduce fuel costs;
- where land is in short supply or expensive;
- where the quality of existing sun dried products can be improved upon;
- where the labour is in short supply;
- where is plenty of sunshine, but high humidity.

(b) Situations where solar dryers may not be useful:

- where conventional energy sources are abundant and cheap;
- where large amounts of combustible by-products or waste materials are freely available;
- where there is insufficient sunshine;
- where is plenty of sunshine and arid conditions (sun drying may suffice);
- where the quality of sun dried products already made cannot be improved upon;
- where local operators are insufficiently trained;
- where the ramifications of introducing a solar dryer have not been completely thought out.

**Sun/solar Drying Tray**

The drying tray described requires seasoned timber 22.5 mm thick x 50 mm wide.

A sun drying tray requires 6 meters of seasoned timber 22.5 mm thick x 50 mm wide. Cut the timber into lengths of 900 mm long for the sides of the tray and 600 mm long for the ends -4 pieces of each length will be needed. The ends of each piece are cut as shown in the drawing—this is to make flush fitting joints. Join the corners using small brass screws 20 mm long. To make extra strong joints use good quality wood glue as well as the screws.

The nylon mosquito netting or grass woven mesh can be fitted between the frames as shown in the bottom drawing. Cut the mesh a little larger than the size of the frame. Using drawing pins, pin the mesh to the outside edges of one of the frames - the mesh should be pulled tight as the pins are put in around the edges.

Lay the other frame on top and drill holes about 3 mm in diameter at the points marked X in the top drawing. Use nails that are a tight fit in the holes and tap gently into place leaving a portion standing above the frame.

Cut off the standing part to leave a piece about 12 mm long which is then bent over and tapped firmly down onto the frame. When the frame has been put together tightly, the drawing pins can be removed.

## DRYERS

### Preservation by Concentration

Foods are concentrated for many of the same reasons that they are dehydrated; concentration can be a form of preservation but this is true only for some foods. Concentration reduces weight and volume and results in immediate economic advantages.

Nearly all liquid foods which are dehydrated are concentrated before they are dried. This is because in the early stages of water removal, moisture can be more economically removed in highly efficient evaporators than in dehydration equipment. Further, increased viscosity from concentration often is needed to prevent liquids from running off drying surfaces or to facilitate foaming or puffing.

Foods are also concentrated because the concentrated forms have become desirable components of diet in their own right. Thus, fruit juices plus sugar with concentration becomes jelly.

The more common concentrated fruit and vegetable products include items as fruit and vegetable juices and nectars, jams and jellies, tomato paste, many types of fruit purees used by bakers, candy makers and other food manufacturers.

**Aspects of Preservation by Concentration**

The level of water in virtually all concentrated foods is in itself more than enough to permit microbial growth. Yet while many concentrated foods such as non-acid fruit and vegetable purees may quickly undergo microbial spoilage unless additionally processed, such items as sugar syrups, jellies and jams are relatively "immune" to spoilage; the difference of course is in what is dissolved in the remaining water and what osmotic concentration is reached.

1. Cell walls
2. Trays on a car
3. Metal plat with burning sulphur
4. Hole for sulphur dioxide fumes
5. Exhaust hole

Tunnel type dehydration unit—tunnel dryer:

1. Control and switch-board
2. Burner
3. Platform for burner
4. Frontal plate
5. Air flow regulating plate
6. Burner's cylinder
7. Air circulating fan
8. Air direction conveying plates
9. Cars rail
10. Tunnel "feeding" door: inlet of cars with fresh product
11. Tunnel "evacuating" door: outlet of cars with dry product
12. Cars pushing device

13. Cars with trays
14. Drying trays

## Typical Counter-flow Tunnel Dryer Construction

Sugar and salt in concentrated solutions have high osmotic pressure. When these are sufficient to draw water from microbial cells or prevent normal diffusion of water into these cells, a preservative condition exists.

The critical concentration of sugar in water to prevent microbial growth will vary depending upon the type of microorganisms and the presence of other food constituents, but usually 70% sucrose in solution will stop growth of all microorganisms in foods.

Less than this concentration may be effective but for short periods of time unless the foods contain acid or they are refrigerated.

Salt becomes a preservative when its concentration is increased and levels of about 18% to 25% in solution generally will prevent all growth of microorganisms in foods. Except in the case of certain briny condiments, however, this level is rarely tolerated in foods. Removal of water by concentration also increases the level of food acids in solution (particularly significant in concentrated fruit juices).

Reduced weight and volume by concentration: While the preservation effects of food concentration are important, the main reason of most food concentration is to reduce food weight and bulk.

Tomato pulp which is ground tomato minus the skins and seeds, has a solid content of only 6 % and so a 3.785 litre can would contain only 231 g of tomato solids.

Concentrated to 32% solids, the same can would contain 1.38 kg of tomato solids or six times the value of product. For a manufacturer needing tomato solids such a producer of soups, canned spaghetti or frozen pizza the saving from concentration are enormous in cans, transportation costs, warehousing costs and handling costs throughout his operation.

### *Specific Gravity and Solids—Tomato Pulp and Commercial Tomato Concentrates*

| *Tomato Solids, % per Gal. at 68°F, lb* | *Specific gravity at 68°F (20°C)* | *Dry Tomato Solids per Litre at 20°C, g* | |
|---|---|---|---|
| 6.0 Tomato pulp | 1.025 | 0.51 | 61 |
| 10.8 | 1.045 | 0.94 | 113 |
| 12.0 Tomato puree | 1.050 | 1.05 | 126 |
| 14.2 | 1.060 | 1.25 | 151 |
| 16.5 | 1.070 | 1.47 | 177 |
| 25.0 | 1.107 | 2.31 | 277 |
| 26.0 | 1.112 | 2.41 | 289 |
| 28.0 Tomato paste | 1.120 | 2.61 | 314 |
| 30.0 | 1.129 | 2.82 | 339 |
| 32.0 | 1.138 | 3.03 | 364 |

***Source:*** Adapted from National Canners Assoc. (1956)

**Methods of Concentration**

a. *Solar Concentration:* As in food dehydration, one of the simplest methods of evaporating water is with solar energy. A typical example of this method is production at farm level in developing countries of fruit pastes/ leathers (such as apricot or plum pastes).

b. *Open Kettles:* Some foods can be satisfactorily concentrated in open kettles that are heated by steam. This is the case for jellies and jams, tomato juices and purees and for certain types of soups. High temperatures and long concentration times should be avoided in order to reduce or eliminate damage. It is also necessary to avoid thickening and burn-on of product to the kettle wall as these gradually lower the efficiency of heat transfer and slow the concentration process.

However, when the process is under control, this type of evaporation is still highly recommended for small scale operations in developing countries. It is a quite widely used system, mainly for jellies, jams and marmalades.

c. Flash evaporators.

d. Thin-film Evaporators.

e. *Vacuum Evaporators:* It is common to construct several vacuumised vessels in series so that the product moves from one vacuum chamber to the next and thereby becomes progressively more concentrated in stages.

With such an arrangement the successive stages are maintained at progressively higher degrees of vacuum, and the hot water vapour produced by the first stage is used to heat the second stage, the vapour from the second stage heats the third stage and so on. In this way maximum use of heat energy is made. Such system is called a multiple effect vacuum evaporator and is illustrated. It is a widely used system for concentrated tomato paste.

f. *Freeze Concentration*: This process has been known for many years and has been applied commercially to orange juice. However, high processing costs due largely to losses of juice occlude [unclear] to the ice crystals, have limited the number of installations to date.

g. Ultrafiltration and reverse Osmosis.

Changes from concentration: Obviously concentration that exposes food to 100° C or higher temperatures for prolonged periods can cause major changes in organoleptic and nutritional properties. Cooked foods and darkening of colour are two of the more common heat induced results which must be kept under control during a well designed process with an efficient evaporator which is still "safe".

Microbial destruction is another type of change that may occur during concentration and will be largely dependent upon temperature. Concentration at a temperature of 100 °C or slightly above will kill many microorganisms but cannot be depended on to destroy bacterial spores. When the food contains acid, such as fruit juices, the kill will be greater but again sterility is unlikely.

On the other hand, when concentration is done under vacuum many bacterial types not only survive the low temperatures but multiply in the concentrating equipment. It is therefore necessary to stop frequently and sanitise low temperature evaporators and where sterile concentrated foods are required, to resort to an additional preservation treatment.

## Chemical Preservation

Many chemicals will kill microorganisms or stop their growth but most of these are not permitted in foods; chemicals that are permitted as food preservatives are listed. Chemical food preservatives are those substances which are added in very low quantities (up to 0.2%) and which do not alter the organoleptic and physico-chemical properties of the foods at or only very little.

Preservation of food products containing chemical food preservatives is usually based on the combined or synergistic activity of several additives, intrinsic product parameters (*e.g.* composition, acidity, water activity) and extrinsic factors (*e.g.* processing temperature, storage atmosphere and temperature).

This approach minimises undesirable changes in product properties and reduces concentration of additives and extent of processing treatments. The concept of combinations of preservatives and treatments to preserve foods is frequently called the hurdle or barrier concept. Combinations of additives and preservatives systems provide unlimited preservation alternatives for applications in food products to meet consumer demands for healthy and safe foods.

Chemical food preservatives are applied to foods as direct additives during processing, or develop by themselves during processes such as fermentation. Certain preservatives have been used either accidentally or intentionally for centuries, and include sodium chloride (common salt), sugar, acids, alcohols and components of smoke. In addition to preservation, these compounds contribute to the quality and identity of the products, and are applied through processing procedures such as salting, curing, fermentation and smoking.

Traditional chemical food preservatives and their use in fruit and vegetable processing technologies could be summarised as follows:

1. Common salt: brined vegetables;
2. Sugars (sucrose, glucose, fructose and syrups):
   (i) Foods preserved by high sugar concentrations: jellies, preserves, syrups, juice concentrates;
   (ii) Interaction of sugar with other ingredients or processes such as drying and heating;
3. Indirect food preservation by sugar in products where fermentation is important (naturally acidified pickles and sauerkraut).

Acidulants and other preservatives formed in or added to fruit and vegetable products are as follows:

***Lactic Acid:*** This acid is the main product of many food fermentations; it is formed by microbial degradation of sugars in products such as sauerkraut and pickles. The acid produced in such fermentations decreases the pH to levels unfavourable for growth of spoilage organisms such as putrefactive anaerobes and butyric-acid-producing bacteria. Yeasts and moulds that can grow at such pH levels can be controlled by the inclusion of other preservatives such as sorbate and benzoate.

***Acetic Acid:*** Acetic acid is a general preservative inhibiting many species of bacteria, yeasts and to a lesser extent moulds. It is also a product of the lactic-acid fermentation, and its preservative action even at identical pH levels is greater than that of lactic acid. The main applications of vinegar (acetic acid) includes products such as pickles, sauces and ketchup.

Other acidulants:

- Malic and tartaric (tartric) acids is used in some countries mainly to acidify and preserve fruit sugar preserves, jams, jellies, etc.
- Citric acid is the main acid found naturally in citrus fruits; it is widely used (in carbonated beverages) and as an acidifying agent of foods because of its unique flavour properties. It has an unlimited acceptable daily intake and is highly soluble in water. It is a less effective antimicrobial agent than other acids.

- Ascorbic acid or vitamin C, its isomer isoascorbic or erythorbic acid and their salts are highly soluble in water and safe to use in foods.

## Commonly Used Lipophilic Acid Food Preservatives

Benzoic acid in the form of its sodium salt, constitutes one of the most common chemical food preservative. Sodium benzoate is a common preservative in acid or acidified foods such as fruit juices, syrups, jams and jellies, sauerkraut, pickles, preserves, fruit cocktails, etc. Yeasts are inhibited by benzoate to a greater extent than are moulds and bacteria.

Sorbic acid is generally considered non toxic and is metabolised; among other common food preservatives the WHO has set the highest acceptable daily intake (25 mg/kg body weight) for sorbic acid. Sorbic acid and its salts are practically tasteless and odourless in foods, when used at reasonable levels (< 0.3 %) and their antimicrobial activity is generally adequate.

Sorbates are used for mould and yeast inhibition in a variety of foods including fruits and vegetables, fruit juices, pickles, sauerkraut, syrups, jellies, jams, preserves, high moisture dehydrated fruits, etc.

Potassium sorbate, a white, fluffy powder, is very soluble in water (over 50%) and when added to acid foods it is hydrolysed to the acid form. Sodium and calcium sorbates also have preservative activities but their application is limited compared to that for the potassium salt, which is employed because of its stability, general ease of preparation and water solubility.

## Gaseous Chemical Food Preservatives

***Sulphur Dioxide and Sulphites:*** Sulphur dioxide ($SO_2$) has been used for many centuries as a fumigant and especially as a wine preservative. It is a colourless, suffocating, pungent-smelling, non-flammable gas and is very soluble in cold water (85 g in 100 ml at 25°C).

Sulphur dioxide and its various sulphites dissolve in water, and at low pH levels yield sulphurous acid, bisulphite and sulphite ions. The various sulphite salts contain 50-68% active sulphur dioxide. A pH dependent equilibrium is formed in water and the proportion of $SO_2$ ions increases with decreasing pH values. At pH values less than 4.0 the antimicrobial activity reaches its maximum.

Sulphur dioxide is used as a gas or in the form of its sulphite, bisulphite and metabisulphite salts which are powders. The gaseous form is produced either by burning Sulphur or by its release from the compressed liquefied form.

Metabisulphite are more stable to oxidation than bisulphites, which in turn show greater stability than sulphites.

The antimicrobial action of sulphur dioxide against yeasts, moulds and bacteria is selective, with some species being more resistant than others.

Sulphur dioxide and sulphites are used in the preservation of a variety of food products. In addition to wines these include dehydrated/dried fruits and vegetables, fruit juices, acid pickles, syrups, semi-processed fruit products, etc. In addition to its antimicrobial effects, sulphur dioxide is added to foods for its antioxidant and reducing properties, and to prevent enzymatic and non-enzymatic browning reactions.

Carbon dioxide ($CO_2$) is a colourless, odourless, non-combustible gas, acidic in odour and flavour. In commercial practice it is sold as a liquid under pressure (58 kg per $cm^3$) or solidified as dry ice.

Carbon dioxide is used as a solid (dry ice) in many countries as a means of low-temperature storage and transportation of food products. Beside keeping the temperature low, as it sublimes, the gaseous $CO_2$ inhibits growth of psychrotrophic microorganisms and prevents spoilage of the food (fruits and vegetables, etc.).

Carbon dioxide is used as a direct additive in the storage of fruits and vegetables. In the controlled/ modified environment storage of fruit and vegetables, the correct combination of $O_2$ and $CO_2$ delays respiration and ripening as well as retarding mould and yeast growth.

The final result is an extended storage of the products for transportation and for consumption during the off-season. The amount of $CO_2$ (5-10%) is determined by factors such as nature of product, variety, climate and extent of storage.

***Chlorine:*** The various forms of chlorine constitute the most widely used chemical sanitiser in the food industry. These chlorine forms include chlorine ($Cl_2$), sodium hypochlorite (NaOCl), calcium hypochlorite ($Ca(OCl)_2$) and chlorine dioxide gas ($ClO_2$).

These compounds are used as water adjuncts in processes such as product washing, transport, and cooling of heat-sterilised cans; in sanitising solutions for equipment surfaces, etc.

Important applications of chlorine and its compounds include disinfection of drinking water and sanitation of food processing equipment.

## General Rules for Chemical Preservation

1. Chemical food preservatives have to be used only at a dosage level which is needed for a normal preservation and not more.
2. "Reconditioning" of chemical preserved food, *e.g.* a new addition of preservative in order to stop a microbiological deterioration already occurred is not recommended.
3. The use of chemical preservatives MUST be strictly limited to those substances which are recognised as being without harmful effects on human beings' health and are accepted by national and international standards and legislation.

Factors which determine/ influence the action of chemical food preservatives:

1. Factors related to the chemical preservatives:
   a. chemical composition;
   b. concentration.
2. Factors related to microorganisms:
   (a) microorganism species; as a general rule it is possible to take the following facts as a basis:

- sulphur dioxide and its derivatives can be considered as an "universal" preservative; they have an antiseptic action on bacteria as well as on yeasts and moulds;
- benzoic acid and its derivatives have a preservative action which is stronger against bacteria than on yeasts and moulds;
- sorbic acid acts on moulds and certain yeast species; in higher dosage levels it acts also on bacteria, except lactic and acetic ones;
- formic acid is more active against yeasts and moulds and less on bacteria.

(b) the initial number of microorganisms in the treated product determines the efficiency of the chemical preservative.

The efficiency is less if the product has been contaminated because of preliminary careless hygienic treatment or an incipient alteration. Therefore, with a low initial number of microorganisms in the product, the preservative dosage level could be reduced.

3. Specific factors related to the product to be preserved:

   a. product chemical composition;

   b. influence of the pH value of the product: the efficiency of the majority of chemical preservatives is higher at lower pH values, *i.e.* when the medium is more acidic.

   c. physical presentation and size which the product is sliced to: the chemical preservative's dispersion in food has an impact on its absorption and diffusion through cell membranes on microorganisms and this determines the preservation effect.

Therefore, the smaller the slicing of the product, the higher the preservative action. Preservative dispersion is slowed down by viscous foods (concentrated fruit juices, etc.)

4. Miscellaneous factors

a. *Temperature:* Chemical preservative dosage level will be established as a function of product temperature and characteristics of the micro-flora;

b. *Time:* At preservative dosage levels in employed in industrial practice, the time period needed in order to obtain a "chemical sterilisation" is a few weeks for benzoic acid and shorter for sulphurous acid.

Usual accepted chemical food preservatives:

***Chemical Food Preservatives***

| Agent | Acceptable Daily intake (mg/Kg body weight) | Commonly used levels (%) |
|---|---|---|
| Lactic acid | No limit | No limit |
| Citric acid | No limit | No limit |
| Acetic acid | No limit | No limit |
| Sodium Diacetate | 15 | 0.3-0.5 |
| Sodium benzoate | 5 | 0.03-0.2 |
| Sodium propionate | 10 | 0.1-0.3 |
| Potassium sorbate | 25 | 0.05-0.2 |
| Methyl paraben | 10 | 0.05-0.1 |
| Sodium nitrite | 0.2 | 0.01-0.02 |
| Sulphur dioxide | 0.7 | 0.005-0.2 |

***Source:*** FDA, 1991

For the purpose of this document, some food products in common usage are summarised as follows:

***Citric Acid:*** fruit juices; jams; other sugar preserves;

***Acetic Acid:*** vegetable pickles; other vegetable products;

***Sodium Benzoate:*** vegetable pickles; preserves; jams; jellies; semi-processed products;

***Sodium Propionate:*** fruits; vegetables;

***Potassium Sorbate:*** fruits; vegetables; pickled products; jams, jellies;

***Methyl Paraben:*** fruit products; pickles; preserves;

***Sulphur Dioxide:*** fruit juices; dried / dehydrated fruits and vegetables; semi-processed products.

## Preservation of Vegetables by Acidification

Food acidification is a means of preventing their deterioration in so far as a non-favourable medium for microorganisms development is created. This acidification can be obtained by two ways: natural acidification and artificial acidification.

### Natural Acidification

This is achieved by a predominant lactic fermentation which assures the preservation based on acidoceno-anabiosys principle; preservation by lactic fermentation is called also biochemical preservation. Throughout recorded history food has been preserved by fermentation. In spite of the introduction of modern preservation methods, lactic acid fermented vegetables still enjoy a great popularity, mainly because of their nutritional and gastronomic qualities.

The various preservation methods discussed thus far, based on the application of heat, removal of water, cold and other principles, all have the common objective of decreasing the number of living organisms in foods or at least holding them in check against further multiplication. Fermentation processes for preservation purposes, in contrast, encourage the multiplication of microorganisms and their metabolic activities in foods. But the organisms that are encouraged are from a select group and their metabolic activities and end products are highly desirable. The extent of this desirability is emphasised by a partial list of fermented fruits and vegetable products from various parts of the world.

There are some characteristic features in the production of fermented vegetables which will be pointed out below using cucumbers as an example.

In the production of lactic acid fermented cucumbers, the raw material is put into a brine without previous heating. Through the effect of salt and oxygen deficiency the cucumber tissues gradually die. At the same time, the semi-permeability of the cell membranes is lost, whereby soluble cell components diffuse into the brine and serve as food substrate for the microorganisms.

Under such specific conditions of the brine the lactic acid bacteria succeed in overcoming the accompanying microorganisms and lactic acid as the main metabolic products is formed. Under favourable conditions (for example moderate salt in the brine, use of starter cultures) it takes at least 3 days until the critical pH value of 4.1 or less-desired for microbiological reasons - is reached.

Beside the typical taste, for the consumer a crisp texture is the most important quality criterion for fermented vegetables. Because there is no heating step before the fermentation, the indigenous plant enzymes in the fermenting materials are still present during the very first phase. After the destruction of the cell membranes they easily get to their active sites and under favourable conditions they can easily cause softening.

The environmental conditions act in a different manner on single enzymes or enzymes systems: some enzymes are strongly inhibited by salt, others are activated, and in the acid pH- region many enzymes are irreversibly inactivated. Beside indigenous enzymes also enzymes produced by microorganisms can be responsible for the undesired soft products.

In technically advanced societies the major importance of fermented foods has come to be variety they add to the diet. However, in many less developed areas of the world, fermentation and natural drying are the major food preservation methods and as such are vital to survival of a large proportion of the world's current population.

Artificial acidification is carried out by adding acetic acid which is the only organic acid harmless for human health and stable in specific working conditions; in this case biological principles of the preservation are acidoanabiosys and, to a lesser extent, acidoabiosys.

Combined acidification is a preservation technology which involves as a preliminary processing step a weak lactic fermentation followed by acidification (vinegar addition).

The two main classes of vegetables preserved by acidification are sauerkraut and pickles; the definitions of these products adapted from US Code of Federal Register (7 CFR 52, 1991) are as follows. Bulk sauerkraut. Bulk or barrelled sauerkraut is the product of characteristic acid flavour, obtained by the full fermentation, chiefly lactic, of properly prepared and shredded cabbage in the presence of 2-3% salt. On completion of fermentation, it contains not more than 1.5% of acid, expressed as lactic acid.

Canned sauerkraut. Canned (or packaged) sauerkraut, is prepared from clean, sound, well-matured heads of the cabbage plant (Brassica oleracea var. capitata L.) which have been properly trimmed and cut; to which salt is added and which is cured by natural fermentation.

The product may or may not be packed with pickled peppers, pimientos, or tomatoes or contain other flavouring ingredients to give the product specific flavour characteristics.

The product:

a) may be canned by processing sufficiently by heat to assure preservation in hermetically sealed containers; or

b) may be packaged in sealed containers and preserved with or without the addition of benzoate of soda or any other ingredient permissible under the provisions of Food and Drug Administration (FDA).

"Pickles" means the product prepared entirely or predominantly from cucumbers (Cucumis sativus L.). Clean, sound ingredients are used which may or may not have been previously subjected to fermentation and curing in a salt brine (solution of sodium chloride, NaCl).

The prepared pickles are packed in a vinegar solution to which may be added salt and other vegetables, nutritive sweeteners, seasonings, flavourings, spices, and other ingredients permissible under FDA regulations. The product is packed in suitable containers and heat treated, or otherwise processed to assure preservation. Sauerkraut and pickle products can be preserved under the effect of natural or added acidity, followed by pasteurization when this acidification is not sufficient.

Sauerkraut is a very good source of vitamin C; the importance of this product should be emphasised in developing countries as a simple technology which can be applied mainly for consumption of the finished products in remote, isolated areas during the cold season.

It is also a excellent technology to be learned to schools which have their own source of cabbage and cucumbers through school agricultural farms.

Sauerkraut and pickles are manufactured on an industrial scale in significant quantities worldwide. However, the basic technology is simple and could be applied at home, farm and community level after some explanation and training. The natural acidification preservation could be considered similar to sun/solar drying in terms of training and development.

### *Some Industrial Fermentation Processes in Food Industries*

| | |
|---|---|
| **I. Lactic acid bacteria** | |
| - cucumbers | dill pickles, sour pickles |
| - cabbage | sauerkraut |
| - turnips | sauerruben |
| - lettuce | lettuce kraut |
| - mixed vegetables, turnips, radish, cabbage | |
| - mixed Chinese vegetables, cabbage | Kimchi |
| - vegetables and milk | Tarhana |
| - vegetables and rice | Sajur asin |
| **II. Lactic acid bacteria with other micro-organisms** | |
| - with yeasts | Nukamiso pickles |
| - with moulds | tempeh, soy sauce |
| **III. Acetic acid bacteria - wine, cider or any alcoholic and sugary or starchy products may be converted to vinegar** | |
| **IV. Yeasts** | |
| - fruit | wine, vermouth |

***Source:*** Pederson (1)

## PRESERVATION WITH SUGAR

The principle of this technology is to add sugar in a quantity that is necessary to augment the osmotic pressure of the product's liquid phase at a level which will prevent microorganism development. From a practical point of view, however, it is usual to partially remove water (by boiling) from the product to be preserved, with the objective of obtaining a higher sugar concentration. In concentrations of 60% in the finished products, the sugar generally assures food preservation.

It is important to know the ratio between the total sugar quantity in the finished product and the total sugar concentration in the liquid phase because this determines, in practice, the sugar preserving action. The percent composition of a product preserved with sugar, for example marmalade, can be expressed as follows: $[i + S + s + n + w] = 100$;

i = insoluble substance;

s = sugar from fruits;

S = added sucrose;

n = soluble "non sugar"

w = water.

In this case, total sugar concentration, in the liquid phase, of the finished product is:

$X = 100 (S + s) / 100 - (n + i) [\%]$

Therefore, in the case of a standard marmalade with 55 % sugar added (calculated on the finished product basis), the real concentration in the liquid phase is for example:

$X = 100 (55 + 8) / 100 - (5 + 3) = 68.5\%$

In the food preservation with sugar, the water activity cannot be reduced below 0.845; this value is sufficient for bacteria and neosmophile yeast inhibition but does not prevent mould attack. For this reason, various means are used to avoid mould development:

- finished product pasteurization (jams, jellies, etc.);
- use of chemical preservatives in order to obtain the antiseptisation of the product surface.

It is very important from a practical point of view to avoid any product contamination after boiling and to assure an hygienic operation of the whole technological process (this will contribute to the prevention of product moulding or fermentation). Storage of the finished products in good conditions can only be achieved by ensuring the above level of water activity.

## HEAT PRESERVATION/HEAT PROCESSING

### Various Degrees of Preservation

There are various degrees of preservation by heating; a few terms have to be identified and understood.

a. *Sterilisation:* By sterilisation we mean complete destruction of microorganisms. Because of the resistance of certain bacterial spores to heat, this frequently means a treatment of at least 121° C (250° F) of wet heat for 15 minutes or its equivalent. It also means that every particle of the food must receive this heat treatment. If a can of food is to be sterilised, then immersing it into a 121° C pressure cooker or retort for the 15 minutes will not be sufficient because of relatively slow rate of heat transfer through the food in the can to the most distant point.

b. *"Commercially Sterile":* Term describes the condition that exists in most of canned or bottled products manufactured under Good Manufacturing Practices procedures and methods; these products generally have a shelf-life of two years or more.

c. Pasteurized means a comparatively low order of heat treatment, generally at a temperature below the boiling point of water. The more general objective of pasteurization is to extend product shelf-life from a microbial and enzymatic point of view; this is the objective when fruit or vegetable juices and certain other foods are pasteurized. Pasteurization is frequently combined with another means of preservation - concentration, chemical, acidification, etc.

d. Blanching is a type of pasteurization usually applied to vegetables mainly to inactivate natural food enzymes. Depending on its severity, blanching will also destroy some microorganisms.

## Determining Heat Treatment/thermal Processing Steps

Since heat sufficient to destroy microorganisms and food enzymes also usually has adverse effects on other properties of foods, in practice the minimum possible heat treatment should be used which can guarantee freedom from pathogens and toxins and give the desired storage life; these aims will determine the choice of heat treatment. In order to safely preserve foods using heat treatment, the following must be known:

(a) what time-temperature combination is required to inactivate the most heat resistant pathogens and spoilage organisms in one particular food?

(b) what are the heat penetration characteristics in one particular food, including the can or container of choice if it is packaged?

***Heat Processing Requirements—Dependence on Product Acidity***

| FRUITS | Drying Conditions | | | Finished Product | |
|---|---|---|---|---|---|
| | Load kg/m² | Temperature °C | Time | Moisture % | Yield % |
| Plums | 15 | I. 40-50 | 6 H | 18-20 | 25-35 |
| | | II. 75-80 | 14 H | | |
| Apples (Rings) | 10 | 75-55 | 5-6 H | 20 | 10-12 |
| Apricots (Halves) | 10 | 70-60 | 10- 15 | 15-20 | 10-15 |
| Cherries (w. stones) | 10 | 55-70 | 6-8 | 12-15 | 25 |
| Pears (Halves and quarters) | 15 | 70-65 | 15-22 | 18-20 | 10-15 |
| | 15 | 70-60 | 10-15 | 15-20 | 10-15 |

*Source:* Desrosier and Desrosier (1977)

Preservation processes must provide the heat treatment which will ensure that the remotest particle of food in a batch or within a container will reach a sufficient temperature, for a sufficient time, to inactivate both the most resistant pathogen and the most resistant spoilage organisms if it is to achieve sterility or "commercial sterility", and to inactivate the most heat resistant pathogen if pasteurization for public health purposes is the goal. Different foods will support growth of different pathogens and different spoilage organisms so the target will vary depending upon the food to be heated.

Food acidity/pH value has a tremendous impact on the target in heat preservation/ processing.

Sequence of operations employed in heat preservation of foods (fruit and vegetables, etc.) In a simplified manner, the main operations employed in heat preservation can be described as follows:

***Food Preparation:*** Preparation procedures will vary with the type of food. For fruit, washing, sorting, grading, peeling, cutting to size, pre-cooking and pulping operations may be employed.

***Can/receptacle:*** This may be carried out manually or by using sophisticated filling machinery. The ratio of liquid to solid in the can must be carefully controlled and the can must not be overfilled. A headspace of 6-9 mm depth (6-8% of the container volume) above the level of food in the can is usual.

***Vacuum Production:*** This can be achieved by filling the heated product into the can, by heating the can and contents after filling, by evacuating the headspace gas in a vacuum chamber, or by injecting superheated steam into the headspace. In each case the can end is seamed on immediately afterwards.

***Thermal Processing:*** The filled sealed can must be heated to a high temperature for a sufficient length of time to ensure the destruction of spoilage microorganisms. This is usually carried out in an autoclave or retort, in an environment of steam under pressure.

***Cooling:*** The processed cans must be cooled in chlorinated water to a temperature of 37°C. At this temperature the heat remaining is sufficient to allow the water droplets on the can to evaporate before labelling and packing.

***Labelling and Packing:*** Labels are applied to the can body, and the cans are then packed into cases.

In principle, all these operations can also be carried out at the farm/community level using the appropriate small scale equipment, preferably only glass jars (*e.g.* no metal cans).

## Technological Principles of Pasteurization

Physical and chemical factors which influence pasteurization process are the following:

a. temperature and time;

b. acidity of the products;

c. air remaining in containers.

***Pasteurization Processes:*** In pasteurising certain acid juices for example, there are two categories of processes:

(a) Low pasteurization where pasteurization time is in the order of minutes and related to the temperature used; two typical temperature/time combinations are as following:

63° C to 65° C over 30 minutes or

75° C over 8 to 10 minutes.

Pasteurization temperature and time will vary according to:

- nature of product; initial degree of contamination;
- pasteurized product storage conditions and shelf life required.

In this first category of pasteurization processes it is possible to define three phases:

- heating to a fixed temperature;
- maintaining this temperature over the established time period (= pasteurization time);
- cooling the pasteurized products: natural (slow) or forced cooling.

(b) Rapid, high or flash pasteurization is characterized by a pasteurization time in the order of seconds and temperatures of about 85° to 90° C or more, depending on holding time. Typical temperature/time combinations are as follows:

88° C (190° F) for 1 minute;

100° C for 12 seconds;

121°C for 2 seconds.

While bacterial destruction is very nearly equivalent in low and in high pasteurization processes, the 121° C/2 seconds treatment give the best quality products in respect of flavour and vitamin retention. Such short holding times, however, require special equipment which is more difficult to design and generally is more expensive than the 63-65 ° C/30 minutes type of processing equipment.

In flash pasteurization the product is heated up rapidly to pasteurization temperature, maintained at this temperature for the required time, then rapidly cooled down to the temperature for filling, which will be performed in aseptic conditions in sterile receptacles. Taking into account the short time and rapid performance of this operation, flash pasteurization can only be achieved in continuous process, using heat exchangers.

Industrial applications of pasteurization process are mainly used as a means of preservation for fruits and vegetable juices and specially for tomato juice.

During pasteurization it is necessary that a sufficient heat quantity is transferred through the receptacle walls; this is in order that the product temperature rises sufficiently to be lethal to microorganisms throughout the product mass.

The most suitable and practical method to speed up thermopenetration is the movement of receptacles during the pasteurization process. Rapid rotation of receptacles around their axis is an efficient means to accelerate heat transfer, because this has the effect, among others of rapidly mixing the contents.

The critical speed of for this movement is generally about 70 rotations per minute (RPM). This enables a more uniform heating of products, reducing heating time and organoleptic degradation. The food processor will employ no less than that heat treatment which gives the necessary degree of microorganism destruction. This is further ensured by periodic inspection by local sanitary authorities or by the importing countries sanitary services. However, the food processor also will want to use the mildest effective heat treatment to ensure highest food quality.

It is convenient to separate heat preservation practices into two broad categories: one involves heating of foods in their final containers, the other employs heat prior to packaging.

The latter category includes methods that are inherently less damaging to food quality, where the food can be readily subdivided (such as liquids) for rapid heat exchange. However, these methods then require packaging under aseptic or nearly aseptic conditions to prevent or at least minimise recontamination.

On the other hand, heating within the package frequently is less costly and produces quite acceptable quality with the majority of foods and most of our present canned food supply is heated in the package.

In practice, therefore, most of the canned food produced locally in developing countries should be heated within the package.

# 4

# METHODS OF PROCESSING

Osmotic dehydration is a useful technique for the concentration of fruit and vegetables, realised by placing the solid food, whole or in pieces, in sugars or salts aqueous solutions of high osmotic pressure. It gives rise to at least two major simultaneous counter-current flows: a significant water flow out of the food into the solution and a transfer of solute from the solution into the food.

***Process Variables:*** Main process variables are:

a. pre-treatments;

b. temperature;

c. nature and concentration of the dehydration solutions;

d. agitation;

e. additives.

In the light of the published literature, some general rules can be noted:

- water loss and solid gain are mainly controlled by the raw material characteristics and are certainly influenced by the possible pre-treatments;
- it is usually not worthwhile to use osmotic dehydration for more than a 50% weight reduction because of the decrease in the osmosis rate over time. Water loss mainly occurs during the first 2 hr and the maximum solid gain within 30 min.
- the rate of mass exchanges increases with temperature but above 45°C enzymatic browning and flavour

deterioration begin to take place. High temperatures, *i.e.* over 60° C, modify the tissue characteristics so favouring impregnation phenomena and thus the solid gain;

- the best processing temperature depends on the food; mass exchanges are favoured by using high concentration solutions;
- phenomena which modify the tissue permeability, such as over-ripeness, pre-treatments with chemicals ($SO_2$), blanching or freezing, favour the solid gain compared to water loss because impregnation phenomena are enhanced;
- the kind of sugars utilised as osmotic substances strongly affects the kinetics of water removal, the solid gain and the equilibrium water content. Low molar mass saccharides (glucose, fructose, sorbitol, etc.) favour the sugar uptake;
- addition of NaCl to osmotic solutions increases the driving force for drying.

Synergistic effects between sugar and salt have also been observed. A pilot plant equipment used for detailed study of process parameters in osmotic concentration of fruits. (Source: Garrote, R.L. et al., 1992).

Applications: The effects of osmotic dehydration as a pre-treatment are mainly related to the improvement of some nutritional, organoleptic and functional properties of the product.

As osmotic dehydration is effective at ambient temperature, heat damage to colour and flavour is minimised and the high concentration of the sugar surrounding fruit and vegetable pieces prevents discoloration.

Furthermore, through the selective enrichment in soluble solids high quality fruit and vegetables are obtained with functional properties "compatible" with different food systems. These effects are obtained with a reduced energy input over traditional drying process. The main energy-consuming step is the reconstitution of the diluted osmotic solution that could be obtained by concentration or by addition of sugar. Various applications of the technique as a unit operation in the food area are summarised in Pig. 8.4.1 together with the process parameters regarded as optimal in the light of the published literature.

Drying: Air drying following osmotic dipping is commonly used in tropical countries for the production of so-called "semi-candied" dried fruits. The sugar uptake, owing to the protective action of the saccharides, limits or avoids the use of SO2 and increases the stability of pigments during processing and subsequent storage period.

The organoleptic qualities of the end product could also be improved because some of the acids are removed from the fruit during the osmotic bath, so a blander and sweeter product than ordinary dried fruits is obtained. Owing to weight and volume reduction, loading of the dryer can be increased 2-3 times.

The combination of osmosis with solar drying has been put forward, mainly for tropical fruit. A 24 hour cycle has been suggested combining osmodehydration, performed during the night, with solar drying during the day. Two-three-fold increase in the throughput of typical solar dryers is feasible, while enhancing the nutritional and organoleptic quality of the fruits.

A two-step drying process, OSMOVAC, for producing low moisture fruit products was described. The osmotic step is performed with sucrose syrup 65-75 Brix until the weight reduction reaches 30-50%.

By osmotic dehydration followed by vacuum drying puffy products with a crisp, honeycomb-like texture can be obtained at a cost comparatively lower than freeze-drying.

Commercial feasibility of the process on bananas has been studied, based on the results of a semi-pilot scale operation; the process scheme is reported. Osmotically dried bananas retained more puffiness and a crisper texture than simple vacuum dried ones, and the flavour lasted longer at ambient temperature.

The combination of osmotic dehydration with freeze-drying has been proposed only at laboratory scale. Appertisation: A combination of osmotic dehydration with appertisation has been proposed to improve canned fruit preserves. The feasibility of a process, called osmo-appertisation, to obtain high quality fruit in syrup, has been assessed on a pilot scale.

The key point of this technique is the pre-concentration of the fruit to about 20-40 Brix, that causes, together with the enhancement of the natural flavour, an increase of the resistance of the fruit to the following heat treatment, especially for colour and texture stability.

The products obtained are stable up to 12 months at ambient temperature and show a higher organoleptic quality than canned preserved alternatives. Furthermore, because of their higher specific weight and diminished volume, the filling capacity of jars or pouches is increased.

***Freezing:*** The frozen fruit and vegetable industry uses much energy in order to freeze the large quantity of water present in fresh products. A reduction in moisture content of the material reduces refrigeration load during freezing.

Other advantages of partially concentrating fruits and vegetables prior to freezing include savings in packaging and distribution costs and achieving higher product quality because of the marked reduction of structural collapse and dripping during thawing.

The products obtained are termed "dehydro-frozen" and the concentration step is generally carried out through conventional air drying, the additional cost of which has to be taken into account.

Osmotic dehydration could be used instead of air drying to obtain an energy saving or a quality improvement especially for fruit and vegetable sensitive to air drying.

## Extraction of Juices

An osmotic pre-step before juice extraction was reported to give highly aromatic fruit or vegetable juice concentrates.

Further developments: So far only applications on a pilot plant scale are reported in the literature. For further developments on a larger scale, theoretical and practical problems should be solved.

The industrial application of the process faces engineering problems related to the movement of great volumes of concentrated sugar solutions and to equipment for continuous operations. The use of highly concentrated sugar solutions creates two major problems.

The syrup's viscosity is so great that agitation is necessary in order to decrease the resistance to the mass transfer on the solution side.

The difference in density between the solution (about 1.3 kg/litre) and fruit and vegetables (about 0.8 kg/litre), makes the product float. Another important aspect, so far not investigated, is the microbiological safety of the process, which should be studied thoroughly before further industrial development.

## Osmoappertisation in the Processing of Apricots

In order to obtain an alternative to the canned fruit preserves and to maintain a high quality of the fruits, a research has been carried out on the osmoappertisation of apricots, a "combined" technique that consists in the appertisation of the osmodehydrated apricots.

This technique could contributes also to the reduction of energy consumption, limits the cost of production and combines "convenience" (ready-to-eat, medium shelf-life) with many market outlets (retail, catering, bakery, confectionery, semi-finished products). Osmoappertisation combines two unit operations: dehydration by osmosis and appertisation (packaging + pasteurization).

## Apricot Processing

*a) Fresh apricot puree.*

After washing, cutting and removal of stones, apricot halves are dipped in 2% solution of sodium or potassium metabisulphite for 10 minutes. After draining, the resulting material is passed through a 0.045-in. screen pulper - finisher to produce a fresh apricot puree.

The fresh apricot puree obtained in this way could be further processed in different semiprocessed (*i.e.* chemically or otherwise preserved products) or finished fruit products (fruit leathers, fruit bars, jams, etc.).

*b) Concentrated apricot pulp.*

Fresh apricot halves could also be steam blanched for 5 min., passed through a 0.045-in pulper - finisher and transformed in a puree with about 14 Brix depending on the fruit quality. This puree may be concentrated in steam jacketed kettles up to 20 Brix or in other adequate equipment (*e.g.* a stirred vacuum evaporator) up to 28°Brix.

As for fresh apricot puree, the concentrate may be further processed in various semiprocessed or finished fruit products as mentioned above and as will be described below.

*c. Dried apricot leather.*

- From fresh fruit puree by drum drying. The fresh apricot puree at about 14 Brix could be dried to 12% moisture apricot sheets, using a double-drum dryer operating at 132 degrees C with a drum clearance of 0.008 in and speed of 45 sec per revolution.
- From fruit concentrate by drum drying. The concentrate could also be dried to 12% moisture fruit sheets by the same process as described above.
- From fresh fruit puree or from apricot concentrate by sun/solar drying or by dehydration.

a) *Trays:* For sun / solar drying or dehydration of fruit pulp, the trays must have a solid base in order to retain the liquid contents. They may be made of metal, timber or plastic. Stainless steel or plastic trays are most suitable because they are unaffected by acid fruit pulp; they are, however, expensive.

A metal tray could be 75 x 50 cm in size and with side 5 cm high. The trays must keep level during drying; if the tray is not level the pulp will run to the lowest point, giving a layer of irregular depth which will dry unevenly. Any tray which is not made of stainless steel or plastic must be covered inside with a sheet of heavy gauge plastic film to protect the pulp from chemical or bacteriological contamination.

Standard sun/solar trays as described can be used by covering them inside with a sheet of plastic film to create a solid base.

b) *Preparation before Drying/dehydration:* Fresh apricot puree can be directly used for next processing steps. Fruit concentrate needs to be added to potassium metabisulphite to obtain a 0.3% concentration of $SO_2$ in the material.

c) *Drying/dehydration:* The apricot/fruit puree or concentrate is poured into the trays to a depth of about 1.5 cm. When stainless steel or plastic trays are used they should be coated with a thin layer of glycerine to prevent sticking.

The pulp is then sun/solar dried or tunnel/cabinet dehydrated; moisture content in the dried product should not exceed 14% and the $SO_2$ content should not be less than 1500 pp.

The dried product is wrapped in cellophane to prevent sticking, then put inside polythene bags and stored at best in tight fitting tins and sealed to prevent moisture transfer.

From fresh fruit puree or from apricot concentrate, with sugar addition, and then processed by sun / solar drying or by dehydration.

In some countries preference is for finished products with added sugar; and this is also interesting from a point of view of energy consumption (concentration is partially achieved by sugar dry matter) and of shelf life. The overall content in $SO_2$ could also be reduced as sugar is a preservation agent, the product will be close to a fruit "paste".

## DRIED/DEHYDRATED PRODUCTS

In reconstitution water is added to the product which is restored to a condition similar to that when it was fresh. This enables the food product to be cooked as if the person was using fresh fruit or vegetable.

All vegetables are cooked but many of the dried fruits can be used for eating after they have been soaked in water. The following reconstitution test is used to find out the quality of the dried product.

**Reconstitution Test**

1. Weigh out a sample of 35 grams from the bulked and packed final product of the previous day's production.
2. Put the sample into a small container (beaker) and add 275 ml of cold water (and 3.5 g salt).
3. Cover the container (with a watch-glass) and bring the water to the boil.
4. Boil GENTLY for 30 minutes.
5. Turn out the sample onto a white dish.
6. At least two people should then examine the sample for palatability, toughness, flavour and presence or absence of bad flavours. The testers should record their results independently.
7. The liquid left in the container should be examined for traces of sand/soil and other foreign matter.

This test can be used also to examine dried products after they have been stored for some time. Evaluation of rehydration ratio may be performed according to the following calculations.

Rehydration ratio. If weight of the dried sample is 10 g (Wd) and the weight of the sample after rehydration is 60 g (Wr), rehydration ratio is:

Rehydration coefficient. The weight of rehydrated sample is 60 g (Wr); the weight of dried sample is 10 g (Wd) and its moisture is 5% (Wu); raw material before drying had 87% water (A); rehydration coefficient is:

A simpler test for eating quality can be carried out without weighing and measuring. The material is placed in a cooking pot with water (and a little salt). The pot is then covered and boiled as described above.

Except for a few products which are eaten in the dry state, most dried fruit and all dried vegetables are prepared by soaking and cooking. Often this preparation is carried out incorrectly and dried products get a bad reputation.

Good quality dried products, after cooking and if properly treated should be similar to cooked fresh produce. In order to get good results, the following methods are recommended:

***Quick Method:*** Cold water, ten times the weight of the dry product, is added to the dried product. The container is covered, brought to the boil and simmered GENTLY until the product is tender. The cooking time may be 15 to 45 minutes after the boiling point has been reached.

***Slow Method:*** This gives better results than the quick method. Cold water is added to the dry food and is left to soak for 1 to 2 hours before cooking. The product is then cooked in the same water as that in which it was soaked. The actual cooking time will probably be shorter than that for the quick method.

Other points to remember are:

- if too much water is added the cooked product will have little flavour. However, if too little water is added the product may dry and burn. This can be avoided by adding small quantities of water during cooking;
- always cook with a lid on the container;
- salt, if required, should be added when the cooking is almost complete;
- partly used packages of dry products should be reclosed tightly or kept in containers with good fitting lids.

***Composition of Some Dried Fruits***

| Fruit | Moisture content % | Sugar as monosaccharides, % | Other carbohydrates | Vitamin C mg/100 g |
|---|---|---|---|---|
| Raisins | 21.5 | 64 | 7 | 0.0 |
| Dates | 12.5 | 55 | 7 | 0.0 |
| Prunes | 20.0 | 45 | 15 | 0.0 |
| Apricots | 15.0 | 45 | 24 | 4.8 |
| Peaches | 15.5 | 55 | 14 | 8.0 |

Handling, sorting, packing and storage of dried and dehydrated fruit and vegetables.

It is not easy to assess when drying has been completed. In absence of instrumentation, the characteristics of the various products after drying/dehydration can only be assessed by experience. Although this cannot be conveyed adequately on paper, some general indications can be given.

***Fruit Products:*** When a handful of fruit is squeezed tightly together in the hand and then released, the individual pieces should drop apart readily and no moisture be left behind on the hand. It should not be possible to separate the skin by rubbing unpeeled fruit and the fruit centre should no longer reveal any moist area. Banana should be leathery and not too tough to eat in their dry state.

***Vegetable Products:*** Onions should be dried until they are crisp whereas tomatoes should be leathery.

In general, the lower the moisture content, the better the keeping quality will be, but overdried products generally have an inferior quality. Also the loss in weight from excessive drying cannot be tolerated in a commercial operation designed to run profitably.

It is, however, essential to dry up to an optimum / safe moisture level, related to the type of the product and his designed shelf life, and to avoid running the risk of the products becoming spoiled due to excess water content. When drying is completed, the material should be sorted either on trays or on a table in order to remove pieces of poor quality and colour and any foreign matter.

Very fine material should be separated from the bulk of the material by using a sieve (12 or 16 mesh per inch). Bad quality products which show poor colour need to be removed from the bulk of finished product. After selection and grading, dried products should be packed immediately, preferably in polythene bags which must be folded and closed / tied tightly. However, plastic bags are easily damaged and therefore they must be packed into cartons or jute sacks before they are transported.

## DETERIORATION OF DRIED FRUIT DURING STORAGE

Dried fruit must be considered as a relatively perishable commodity in the same category as cereals, pulses and similar stored products. It is subject to deterioration resulting from mould growth, insect and mite infestation and physical and chemical changes.

### Mould Growth

When the moisture content of dried fruit is allowed to exceed the maximum permissible level for safe storage then mould growth may occur. The moisture levels applicable to various types of fruit; and it can be seen that the safe moisture levels for dried fruit are much higher than those for other similar commodities.

At the present time, suitable field moisture meters for use with dried fruit are not readily available, and moisture determinations can only be satisfactorily carried out where laboratory facilities are available.

Various species of drought resisting fungi may develop on dried fruit when the moisture content is just above the safe level, and a number of osmophilic yeasts are quite commonly associated with spoilage in dried fruit.

Many of the yeasts bring about fermentation with the production of lactic acid or alcohol, and yeasts are frequently present in wart-like crystalline growths which occur in fruit which has become "sugared". In very moist fruit mucoraceous fungi may predominate and are visible as white fluffy growths on and within the fruit.

***Mite Infestation:*** Severe mite infestations are often associated with the growth of osmophilic yeasts in fermenting dried fruit products.

Many of these mites are unable to complete their development in the absence of yeast. They have been reported as occurring on dried fruit, and particularly figs and prunes in Mediterranean countries. Such infestations are difficult to eradicate and affect consumer acceptance of the contaminated products.

***Insect Infestation:*** Insect infestations may begin in the field before harvest, may continue during bulk storage after drying, and unless measures are taken to prevent it, may occur in the finished packaged product during storage prior to distribution and consumption.

Regular treatments of the stack of dried fruit with a suitable insecticide will be necessary as a routine to combat light insect infestations. Pyrethrins synergised with piperonyl butoxide are commonly used as a surface spray or as an aerosol fog for this purpose. Heavy infestations will require that the fruit be fumigated.

## TECHNOLOGY OF SEMI-PROCESSED FRUIT PRODUCTS

The semi-processed fruit products are manufactured in order to be delivered to industry processing centres (in the fruit producing country itself or in importing countries) where they will be further manufactured in consumer oriented finished products: jams, jellies, syrups, fruits in syrup, etc.

In the practice of semi-processed fruit products and for the purpose of this document the following categories are defined:

a. *Fruit "Pulps":* semi-processed products, not refined, obtained by mechanical treatment (or, less often, by thermal treatment) of fruit followed by their preservation. Either whole fruit, halves or big pieces are used which enables easy identification of the species. "Pulps" can be classified in boiled or non boiled (raw).

b. *Fruit "Purees-marks":* semi-processed products obtained by thermal and mechanical treatment or, very rare, raw and then refined, operations by which all nonedible parts (cores, peels, etc.) are removed. "Purees-marks" are classified in boiled (the more usual case) and non boiled (raw).

c. *Semi-processed Juices:* products obtained by cold pressure or very rare by other treatments (diffusion, extraction, etc.) followed by the preservation.

## Technical Processes for Preservation of Semi-processed Fruit Products

Preservation can be achieved by chemical means, by freezing or by pasteurization. The choice of preservation process for each individual case is a function of the semi-processed product type and the shelf life needed.

***Chemical Preservation:*** In many countries, in practice, this is carried out with sulphur dioxide, sodium benzoate, formic acid and, on a small scale, with sorbic acid and sorbates.

Preservation with sulphur dioxide is a widespread process because of its advantages: universal antiseptic action and very economic application. The drawbacks of $SO_2$ are: $SO_2$ turn firms the texture of some fruit species (pomaces), desulphiting is not always complete and recolouring of red fruits is not always complete after desulphitation.

Practical preservation dosage levels with $SO_2$ for about 12 months is 0.18-0.20% $SO_2$ (with respect to the product to be preserved).

This level could be reduced to 0.09% $SO_2$ for 3 months and to 0.12% $SO_2$ for 6 months preservation.

The preservation with sulphur dioxide is in use mainly for "pulps" and for "purees-marks".

Chemical preservation can be performed from a practical point of view by the utilisation of 6% $SO_2$ water solutions or by direct introduction of sulphur dioxide gas in the product (for "purees-marks"). The preparation of 6% $SO_2$ solutions is done by bubbling the gas from cylinders in cold water; from a 50 kg SO2 compressed gas cylinder results 830 l of 6% $SO_2$ solution.

These $SO_2$ solutions have to be stored in cool places, in closed receptacles and with periodic concentration control/check by titration or by density measurements approximate results.

Preservation with sodium benzoate has the following advantages: it does not firm up the texture and does not modify fruit colour. The disadvantages are: it is not a universal antiseptic, its action needs an acid medium and the removal is partial. Sodium benzoate is in use for "pulps" and for "purees-marks" but less for fruit juices. Practical dosage level for 12 months preservation is 0.18-0.20 % sodium benzoate, depending on the product to be preserved. Sodium benzoate is used as a solution in warm water; the dissolution water level has to be at maximum 10% reported to semi-processed product weight.

Formic acid preservation is performed mainly for semi-processed fruit juices at a dosage level of 0.2 % pure formic acid (100%). Formic acid is an antiseptic effective against yeasts, does not influence colour of products and is easily removed by boiling.

Formic acid could be diluted with water in order to insure a homogeneous distribution in the product to be preserved; water has to be at maximum 5 % of the product weight. Because of a potential effect of pectic substance degradation, formic acid is less in use for "pulps" and "purees-marks" preservation.

Sorbic acid used as potassium sorbate (easily water soluble) can be used for preservation of fruit semi-processed products at a dosage level of 0. 1% maximum. Advantages of sorbates are: they are completely harmless and without any influence on the organoleptic properties of semi-processed fruit products.

***Preservation by Pasteurization:*** As fruit has a low pH, preservation of semiprocessed fruit products could also be performed by pasteurization (heat treatment step at maximum temperature of 100° C), the length of this step varying with the size of the receptacles.

The advantages of this type of treatment are: hygienic process, which assure a long term preservation; the disadvantages are: need for air tight receptacles, and pectic substances could begin to deteriorate if the thermal treatment is too long.

Thermal preservation of fruit semi-processed products could also be done by a "selfpasteurization": very hot semi-processed products are filled into receptacles (*e.g.* metal cans) which are sealed and then inverted in order to sterilise the air which goes through the hot fruit mass.

***Preservation by Freezing:*** This is done on an industrial scale in some countries and can be done with or without sugar addition.

The advantages of this process are: absence of added substances; very good preservation of quality of fruit constituents (pectic substances, vitamins, etc.) and good preservation of organoleptic properties (flavour, taste, colour). Freezing is done at about -20 to -30° C and storage at -10 to -18° C.

Freezing is applied mainly to semi-processed fruit products aimed at very high quality and high cost finished products.

Technological flow-sheet for semi-processed fruit "pulps": chemical preservation.

Sorting is needed in order to remove sub-standard fruit (with moulds, with diseases, etc.) and all foreign bodies.

Washing is obligatory in order to remove all impurities which cannot be eliminated at the processing step in finished products.

Coring and Cutting, mainly for pomace fruits, has as main objective a better utilisation of preservation "space" in receptacles and is not mandatory; this will be defined by customer/ supplier agreements / standards. This operation is preferably performed by mechanical means.

Preservation is carried out with the 6% $SO_2$ solution which is added to the prepared fruits (placed in bulk in receptacles) in the quantity needed to obtain the preservation dosage level. For a better / homogeneous preservative distribution, the initial 6% SO2 solution could be diluted with water; however, the diluted solution (which will be filled in receptacles) has to be at a dosage level of less than 10% of the semi-processed product weight.

For some soft fruit, especially strawberries, preservation is done with a mix of 6% SO2 solution and calcium bisulphite solution (containing also 6% $SO_2$).

Preparation of calcium bisulphite solution is done by the introduction of 30 kg of CaO in 1 $m^3$ $SO_2$ solution and mixing up to clarification. The resulting solution is mixed with the initial 6% $SO_2$ solution, generally in a 1:1 ratio, but the ratio can be adapted to the fresh fruit texture. Firming of soft fruit texture by this treatment is based on the formation of calcium pectate with pectic substances from fruit tissues.

In the case of sodium benzoate, formic acid or potassium sorbate, the dosage levels to be used are as indicated above with the rule that it is not allowed to add more that 10% liquid in receptacles on the prepared fruits.

Preservation by pasteurization or "self-pasteurization" will need as additional steps: (a) boiling with a minimum water addition (maximum 10%); (b) filling of receptacles; (c) hermetic closing followed by (d) pasteurization or "self-pasteurization".

Some general technical data for the preparation of chemical preserved semi-processed fruit "pulps".

***General Technical Processing Data for Semi-processed Fruit "Pulps"***

| Fruit species | Preliminary operations | Preservation means |
|---|---|---|
| Apples, pears, quinces | Sorting, washing, coring, cutting | Sulphur dioxide |
| Prunes, peaches, wax cherries, apricots | Sorting, washing, stone removal (pitting) | Sulphur dioxide or sodium benzoate |
| Cherries | Sorting, washing | Sulphur dioxide or sodium benzoate, sometimes with calcium bisulphite addition |
| Strawberries | Sorting, washing | Sulphur dioxide in mix with calcium bisulphite |
| Wild berries | Sorting, washing | Sulphur dioxide or sodium benzoate; in some cases with calcium bisulphite addition |

## Technological Flow-sheet for Semi-processed "Puree-Marks"

The general technological flow-sheet includes the following operations: Sorting and Washing are obligatory and are carried out in a similar manner as for "pulps".

Heat Treatment/Boiling is needed in order to soften the fruit tissues before refining. For some fruits as strawberry and wild berries, this step is not done and fruits are refined "raw" in order to preserve their flavour. Pulping is performed with specific equipment - refiners or pulpers - which eliminate seeds, pits and other non edible parts (peels, cores, etc.). Preservation is carried out by chemical means, by freezing or by pasteurization.

The general technical / processing data for the manufacture of "puree-marks" are seen in Table.

### *General Technical Processing Data for Semi-processed "Pureesper—Marks"*

| Fruit species | Preliminary operations | Preservation means |
|---|---|---|
| Apples, pears, quinces | Sorting, washing, boiling, refining | Sulphur dioxide; in less frequent cases formic acid or sodium benzoate |
| Prunes, peaches, wax cherries, apricots, | Sorting, washing, boiling, refining | Sulphur dioxide, formic acid or sodium benzoate; cherries freezing with or without sugar; self-pasteurization |
| Strawberry, wild berries | Sorting, washing, refining | Chemical preservation, freezing or self pasteurization |

### *Correlation between Density and Concentration for $SO_2$ Solutions*

| Density at 15°C | Concentration of solutions % $SO_2$ |
|---|---|
| 1.0168 | 3.00 |
| 1.0181 | 3.25 |
| 1.0194 | 3.50 |
| 1.0206 | 3.75 |
| 1.0221 | 4.00 |
| 1.0234 | 4.25 |
| 1.0248 | 4.50 |
| 1.0261 | 4.75 |
| 1.0275 | 5.00 |
| 1.0289 | 5.25 |
| 1.0302 | 5.50 |
| 1.0315 | 5.75 |
| 1.0340 | 6.25 |
| 1.0353 | 6.50 |
| 1.0365 | 6.75 |
| 1.0377 | 7.00 |

## FRUIT SUGAR PRESERVES TECHNOLOGY

As a overall rule of thumb, a sugar concentration of about 60% in finished or processed fruit products generally insures their preservation. Preservation is not only determined by the osmotic pressure of sugar solutions but also by the water activity values in the liquid phase, which can be lowered by sugar addition; and by evaporation down to 0.848 aw; this value however does not protect products from mould and osmophile yeast attack. Maximum saccharose concentration that can be achieved in the liquid phase of the product is 67.89%; however higher total sugar quantities (up to 70-72%) found in products are explained by an increased reducing sugar solubility resulting from saccharose inversion.

### Jams

The preservation of fruit by jam making is a familiar process carried out on a small scale by housewives in many parts of the world. Factory jam making has become a highly complex operation, where strict quality control procedures are employed to ensure a uniform product, but the manufacturing operations employed are in essence the same as those employed in the house.

Fresh or pre-cooked fruit is boiled with a solution of cane or beet sugar until sufficient water has been evaporated to give a mixture which will set to a gel on cooling and which contains 32-34% water.

Gel formation is dependent on the presence in the fruit of the carbohydrate pectin, which at a pH of 3.2 - 3.4 and in the presence of a high concentration of sugar, has the property of forming a viscous semi-solid.

During jam boiling, all microorganisms are destroyed within the product, and if it is filled hot into clean receptacles which are subsequently sealed, and then inverted so that the hot jam contacts the lid surface, spoilage by microorganisms will not take place during storage.

The composition of jam made from stone fruit and berry fruit. About 30% of the vitamin C present in fresh fruit is destroyed during the jam-making process, but that which remains in the finished product is stable during storage.

The high moisture content of jam (equivalent to an equilibrium relative humidity of about 82%) makes it susceptible to mould damage once the receptacle has been opened and exposed from some time to the air. No problems of microbiological spoilage are likely to arise in the canned product during storage.

***Composition of Some Fruit Jams***

| Type of Jam | Moisture content % | Sugar (as invert sugar, %) | Vitamin C mg/100 g |
|---|---|---|---|
| Jam made from berry fruits: strawberry, raspberry, etc. | 29.8 | 69.0 | 10 - 25 |
| Jam made from stone fruits: apricot, peach, etc. | 29.6 | 69.3 | 10 - 35 |

***Source:*** FAD/WFP, 1970

## Marmalade

This sugar preserve is defined as "semisolid or gel-like product prepared from fruit ingredients together with one or more sweetening ingredients and may contains suitable food acids and food pectins; the ingredients are concentrated by cooking to such a point that the TSS—Total Soluble Solids- of the finished marmalade is not below 65%".

## Fruit Paste

Fruit paste is a product obtained in the same way as special non-gelified fruit marmalade but with a lower water content-about 25% TSS in fruit paste.

Lowering water content could be achieved by continuing boiling of the product or by drying the product by natural or artificial drying. An example of paste without sugar is the sun dried apricot or prune paste.

General procedure for the preparation of jams, jellies and marmalade

a. Boil the pulp or the juice (with water when necessary)

b. Add the pectin

* to the batch while stirring very vigorously
* Pectin which has previously been mixed with 5 times its weight in sugar taken from the recipe)

c. Boil for about 2 minutes to assure a complete dissolution

d. Add the sugar while keeping the batch boiling

e. Boil down quickly to desired Brix

f. Add the acid (usually citric acid) and remove the froth

g. Fill hot into the (previously cleaned) jars and close

h. Invert the jars for three minutes to pasteurize the cover

* ***Note:*** the pectin in solution can also be added at the end of the step (e) and has to be prepared as follows: use a strong blender. For one litre of water add slowly into the blender 25 g of pectin mixed with 100 g of sugar taken from the recipe.

***Basic Recipes:*** The following recipes must be considered only as guidelines because the composition of the fruit can vary; also the taste of the consumers varies concerning the consistency, the sweetness and acidity.

Before starting to make jam it is important to know the yield to settle the question on containers. The calculation is made as follows:

$$\frac{\text{Total Soluble Solids content of all the raw ingredients}}{\text{Percentage of Total Soluble Solids in the final product}} \times 100$$

In these basic recipes it is assumed that the fruits are poor in pectin content.

Recipe 1. Fruit: sugar = 50:50; desired Brix in the finished product is 68.

| | *Soluble Solids* |
|---|---|
| 10 kg of fruit at 10% TSS | 1.000 kg |
| 10 kg of sugar | 10.000 kg |
| 60 g of pectin (grade 200) | 0.060 kg |
| 55 g of citric acid | 0.055 kg |
| | 11.115 kg |

Recipe 2. Fruit: sugar = 45:55; desired Brix in the finished product is 68.

| | *Soluble Solids* |
|---|---|
| d10 kg of fruit at 10% TSS | 1.000 kg |
| 2.5 litre of water | — |
| 12.2 kg of sugar | 12.200 kg |
| 65 g of pectin grade 200 | 0.060 kg |
| 60 g of citric acid | 0.060 kg |
| | 13.325 kg |

Recipe 3. Fruit: sugar = 40:60; desired Brix in the finished product is 68.

| | *Soluble Solids* |
|---|---|
| d10 kg of fruit at 10% TSS | 1.000 kg |
| 3.3 litre of water | — |
| 15 kg of sugar | 15.000 kg |
| 85 g of pectin grade 200 | 0.085 kg |
| 80 g of citric acid | 0.080 kg |
| | .16.165 kg |

**Processing of Pineapple-papaya Jam**

The fruit should be prepared as per previous instructions.

For pineapples, the ends are removed and discarded; the cores and outer parts of the fruits are also removed. The fruit cylinders obtained are pulped through a special extractor (Fitzpatrick communiting machine) equipped with a 0.40-in screen sieve; the pulp thus obtained is used for making jam.

The papaya are prepared by hand-peeling the fruit; the fruit is then halved and the seeds removed. It is then pulped in the communiting machine using a 0.40-in screen sieve.

When ginger root is used as flavouring, it is peeled and macerated in a Kenwood blender to a very fine consistency.

A typical formula for a pineapple-papaya jam (50:50 ratio) with ginger flavouring is given as follows:

| | |
|---|---|
| Pineapple pulp | 25.0 |
| Papaya pulp | 25.0 pounds |
| Cane sugar | 50.0 |
| Apple pectin (150 grade) | 6.0 ounces |
| Citric acid | 6.4 |
| Fresh ground ginger | 7.5 |

Processing is carried out in the following way:

The weighed fruit pulp is placed in a stainless steel steam-jacketed kettle and heated to about 110°F under constant stirring.

When the product reaches this temperature, the heat is turned off. The pectin (mixed in about ten times its weight with some of the weighed sugar), is then mixed into the fruit pulp, stirring constantly in order to prevent the pectin from clotting.

When the pectin has dissolved, the remainder of the sugar is added and dissolved completely in the mixture. The heat is then turned on and the jam mixture is stirred constantly until it starts boiling vigorously. During the remainder of the cooking, the product is stirred occasionally. Near the finishing point (approximately 221° F), the citric acid and the ginger (if it is used) are also added.

Determination of the finishing point is done by removing samples at intervals, cooling, and reading the soluble solids by means of a refractometer equipped with a Brix scale. After the jam reaches the proper Total Soluble Solids content, the heat is turned off and the surface scum/foam is removed.

The jam then is quickly put into receptacles which have been cleaned and sterilised with boiling in water for 30 minutes. The filling operation is done rapidly in order to prevent the temperature of the jam from falling below 190° F.

After filling, sterilised lids (boiled for 30 minutes in water) are placed on the receptacles and they are then sealed.

After this operation the receptacles are inverted for about 3 minutes to insure that the lids are sterilised. The receptacles are then placed upright. At this stage it is not necessary to do any further processing, therefore the receptacles are cooled in running cold water until they reach a temperature slightly above room temperature. They are then dried in air and labelled.

## Evaluation of Finished Products

During production at medium/large scale, it is recommended that quality controls be performed during manufacturing. After ten weeks of storage at room temperature it is recommended that an examination of finished products be performed. The receptacles are opened and contents carefully emptied on to enamel trays without disturbing the formation of the jam.

The empty cans (if metal cans were used) are then inspected for signs of corrosion. Factors other than flavour include colour, appearance, syrup separation, firmness and spreading quality. For flavour, jam is tested on pieces of bread. Samples are taken for measurement of pH (with a glass electrode pH meter) and Total Soluble Solids (with a refractometer equipped with a Brix scale).

This evaluation enables to have a quality check during product shelf life and to obtain data needed for necessary improvements of future productions.

For pineapple-papaya jam, products made with 30% pineapple and 70% papaya with added ginger has the highest score for flavour. The use of plain tin cans causes corrosion problems which is not the case when acid resistant lacquer cans are used.

## Pineapple Jam Making

1. Boil 40 lb. of pulp and 12 lb. of water.
2. Add 225 g of pectin to the batch while stirring rapidly.
3. Boil for about 90 sec to assure complete dissolution.
4. Add 60 lb. of sugar gradually if possible in several portions, while keeping the batch boiling.
5. Boil down quickly to 69 deg. Brix (223 deg. F).
6. Take off steam, remove foam.
7. Add 300 cc citric acid solution 50%.

8. Fill hot (180 deg. F).
9. Invert receptacles for 3 minutes.

Check each batch for Brix 68-70 deg.; acidity/pH = 3.2 +/- 0.2.

***Evaluate:*** absence of defects; colour; flavour; consistency.

Various fruit:sugar ratios can be manufactured; some basic recipes are as follows:

| | Ratio 50:50 | Ratio 45:55 | Ratio 40:60 |
|---|---|---|---|
| Fruit | 55 lb | 49.5 lb | 44 lb |
| Water | - | 11 lb | 13.2 lb |
| Sugar | 55 lb | 60.5 lb | 66 lb |
| Pectin (150 grade) | 225 g | 237.5 g | 250 g |
| Citric acid (50% sol.) | 300 cc | 320 cc | 335 cc |

# 5

# TECHNOLOGY FOR FRUIT JELLIES

Jellies are gelified products obtained by boiling fruit juices with sugar, with or without the addition of pectin and food acids. Jellies are usually manufactured from juices obtained from a single fruit species only, obtained by boiling in order to extract as much soluble pectin as possible.

Jellies have to be clear, shiny, transparent and with a colour specific to the fruit from which they are obtained. Once the product is removed from the glass receptacles where it was packed, jellies must keep their shape and gelification and not flow, without being sticky or of a too hard consistency.

Technological flow-sheet for jellies manufacturing covers two categories of operations: those to obtain gelifying juices and those related to the manufacturing of jelly itself.

(a) Production of gelifying juices:

Washing & Sorting are carried out in usual conditions;

Cutting is applied eventually only to pomaces (apples, quinces) and are limited to cutting in halves or quarters;

Boiling Is Performed With Water Addition With 50-100% water, needed for pectin extraction. The boiling time is 30-60 min., it should not be longer so as to avoid pectin degradation; at the same time boiling must not be too violent.

Juice Separation is carried out by a simple drain through metallic sieve or cloths; in these cases the yield is lower and the residue can be used for marmalade production. In bigger productions it juice separation by hydraulic press is preferred, yield being in these cases greater.

Juice Clarification is strictly necessary in order to obtain clear jellies. This step can be achieved by sedimentation during 24 hours or by filtration.

(b) Manufacturing of jellies

Basic Recipe Setting is done starting with equal parts in weight of sugar and juice (for example 1000 g juice and 1000 g sugar). As final jelly has to contain about 60% added sugar, weight of finished product must be of about 1600 g, by evaporation of about 400 g water.

Boiling is carried out as following: juice is boiled up to removal of about half of the water that has to be evaporated, then the calculated sugar quantity is added gradually; the remainder of the water is evaporated until a concentration in soluble substances (refractometric extract) of 65-67% is reached, in which is incorporated also the sugar from juice.

During boiling it is necessary to remove foam / scum formed. Product acidity must be brought to about 1% (malic acid) corresponding to pH > 3. Any acid addition is performed always at the end of boiling.

For juices rich in pectin, gelification will occur without pectin addition. If at the trial boiling test the gelification has not occurred, because of pectin absence, in this case 1-2% powder pectin will be added by operating as indicated: pectin is mixed with 10-20 fold sugar quantity and is introduced directly in the partially evaporated juice and then boiling is conducted rapidly up to final point. Evaluation of final point is done not only by refractometry but also by gelifying test.

A rapid test for evaluation of juice pectin content is possible by mixing a small sample of juice with an equal volume of 96% alcohol; the apparition of a compact gelatinous precipitate indicates a sufficient pectin content for gelification.

Boiling of jellies is performed in small batches (25-75 kg) in order to avoid excessively long boiling time which brings about pectin degradation.

Cooling is optional and is carried out up to 85 °C, in double wall baths with water circulation.

Filling is performed at a temperature not below 85 °C in receptacles (glass jars, etc.), which must be maintained still about 24 hours to allow cooling and product gelification.

Receptacle Closing is done after product gelification.

Usual jelly types are: quinces, strawberries, cherries, wild berries, alone or in mixes with apples.

## Grading of Marmalades

Three categories can be defined:

- fine marmalade, manufactured from one fruit;
- superior marmalade, obtained from a mix of fruit in which 30% are "noble" species (cherries, strawberries, apricots, etc.) and 70% from other species;
- marmalade from fruit mixes; apples, pears, plums, quinces, ungrafted apricots and wax cherries may be used, with the optional addition of "superior" fruit which was rejected at sorting but which was sound.

The content in total soluble substances (refractometric extract) of marmalades must be 64% minimum; the acidity must be between 0.5% and 1.8% expressed as malic acid.

Basic Recipe Setting for a normal composition - marmalade without pectin addition the following is a basic recipe:

100 kg semi-processed fruit product:

(10% refractometric extract) ...10 kg soluble substances

55 kg sugar ............................55 kg soluble substances

155 kg 65 kg soluble substances

55 kg water to be evaporated

100 kg marmalade with 65% refractometric extract

This marmalade satisfies many standards and at same time has a good shelf-life since it contains less than 35% water. Semi-processed fruit products must have a minimum 8% refractometric extract; in this case the recipe should use

125 kg of raw material, with 80 kg water to be evaporate. The use of semi-processed fruit products with a low refractometric extract presents the following drawbacks:

a. higher water quantity to be evaporated;

b. longer boiling times with negative impact of pectin degradation;

c. loss of flavour and

d. lower equipment efficiency.

Pectin addition in marmalade manufacture produces the following advantages:

a. improvement of gelification,

b. economy in fruit;

c. shorter boiling time; this maintains taste and flavour and produces higher equipment efficiency.

Pectin addition makes it possible to obtain the "fine" type of marmalade from "noble" fruits which do not contain enough pectin (cherries, peaches, apricots, etc.). In marmalades from fruit mixes, low pectin content can be compensated by addition of semi-processed fruit products which are rich in this component (for example apples).

When pectin is to be added, the above recipe should be modified as follows:

80 kg semi-processed fruit product

(10% refractometric extract) ...8 kg soluble substances

55 kg sugar ........................55 kg soluble substances

10 kg pectic extract (10 % R.E.) ......1 kg soluble substances

145 kg 64 kg soluble substances

45 kg water to be evaporated

100 kg marmalade with 64% refractometric extract

Pectin can be added as pectic extract with about 10% refractometric extract (R.E.) in the recommended proportion or in the form of a powder considered with 100% dry matter (*e.g.* 100% soluble substances) in a quantity of about 1%.

In some countries it is usual to add 5-12% of corn syrup (calculated to finished product weight) to replace the corresponding quantity of sugar (100 parts corn syrup can replace 80 parts sugar). Corn syrup has to be liquefied by heating before use. Corn syrup addition reduces the too sweet taste of marmalade, avoids sugar crystallisation and gives a special shine to finished products.

Marmalade manufacturing follows the technological line drawing and covers the following steps:

"Mark" Preparation can be achieved from fresh fruits (as indicated in the section related to semi-processed fruit products) or starting from chemical preserved semiprocessed fruit products: "marks" or "pulps". In the latter case, pulps will be processed in marks which then will be desulphitated.

Desulphitation is carried out by boiling at atmospheric pressure, under vacuum in specialised equipment or under pressure in special retorts built in acid-resistant material. In any case, the desulphitation must be carried out before sugar addition because sugar will bound to the sulphur dioxide. The desulphitation operation must be conducted so as to be, if possible, fully completed; the finished product must contain less than 0.005% free $SO_2$.

The technological flow for marmalade production according to the following: fresh fruit after sorting on control belt (1) is washed in a washing machine (2) is brought to the continuous boiling equipment (3) then to the pulper (4); the semi-finished mark is passed on to the storage tank (6). Pulps are boiled and desulphited in continuous boiling equipment (3), then are brought to pulper (4) and to the storage tank (6).

Boiling aims at evaporating the required water quantity, to facilitate the formation of pectin-sugar-acid gel and to partially invert sugar (about 40% from total sugar). The boiling operation can be carried out in open kettles or in evaporators under vacuum. In the latter case, the warm "mark" from storage tank (6) is aspirated in a concentrator (7) in a vacuum and submitted to a partial boiling up to removal of half of water quantity which needs to be evaporated; the calculated sugar quantity is then added by aspiration, keeping the boiling on. After this the pectin extract or powder pectin which has previously been dissolved in warm water, is added; when the final concentration is reached, as indicated by refractometric control, the required quantity of acid is added. Sugar is added in proportion of 55% in finished product, pectic extract (10% refractometric extract) at a level of about 10-15% and the acid (citric, tartaric, lactic) in a quantity needed to obtain a finished product acidity of about 1%.

Boiling at atmospheric pressure affects not only the appearance but also the nutritional value of the products, mainly if these contain proteins, as some albuminoids coagulate even at 60°C.

Food products for which flavour is an essential property as for example fruit juices, etc., are also affected by the action of heat. Heat treatment has an impact on vitamin losses, mainly of vitamin C, in the presence of oxygen as is the case at concentration in open vessels.

**Technological Line for Marmalade Production**

Sugars are generally less damaged by heat at temperatures below 100° C; as the boiling point is increasing above 100° C, a risk of partial sugar caramelization exists.

In the study of heat effects on products submitted to concentration operation, it is necessary to take into account not only the evaporation surface temperature but also the distribution of the temperature in the whole liquid mass.

The length of the heating period also has a major influence because in many cases it is preferable to concentrate the liquid at a relatively high temperature in a short time avoiding the drawbacks of lower temperatures acting during a long time.

In order to maintain the food value and organoleptic properties, it is necessary that concentration take place at a low temperature which can be achieved by concentration under partial vacuum, taking into account that boiling point decreases when the residual pressure decreases, respectively with the increase of vacuum degree.

Advantage of concentration under partial vacuum are the following:

- lowering of boiling point;
- the total time needed for concentration of food products under a residual pressure of about 200 mm Hg is about half as compared to the that of concentration by boiling at atmospheric pressure;
- by lowering the concentration temperature and time, organoleptic properties and of nutritional value are maintained better particularly as far as the vitamins are concerned;
- when products are concentrated in a vacuum, it is possible to recover volatile aromatic substances by using adequate installations.

Technical procedures of concentration by vaporisation can be classified in:

a) concentration at atmospheric pressure: continuous or discontinuous;

b) concentration under partial vacuum: discontinuous (in vacuum equipment with simple or multiple effect) or continuous (in vacuum installations with continuous action or in thin film vaporisation installations).

Even if open kettle equipment is less expensive than evaporators in a vacuum, it is necessary to take into account that boiling under vacuum has the following advantages:

a) low boiling temperature (60-70° C), depending the degree of the vacuum; this give the fruit better taste and flavour-keeping qualities;

b) easy feeding with raw and auxiliary materials;

c) shorter boiling time;

d) better working conditions (vapour elimination in condensed water and not in open air).

There are small size evaporators under vacuum which can be well suited to the needs of medium size operations in developing countries. Cooling of marmalade to about 50-60°C can even be done in a vacuum evaporator by closing the heating steam and maintaining vacuum degree or by discharge in storage tanks.

Filling in receptacles (boxes, jars, glasses, etc.) is done preferably with filling machines followed by labelling. Small packages can be closed warm or after complete cooling; big packages (boxes, etc.) must be closed only after cooling, *e.g.* 24 hours after processing.

Storage of marmalade must be done in dry rooms (air relative humidity at about 75%), well ventilated, medium cool places (temperature 10-20 degrees C), disinfected and away from direct sunlight and heat. These measures are necessary because marmalade is a hydroscopic product and, by water absorption, favourable conditions for mould development are created.

## TECHNOLOGY OF SPECIAL FRUIT JAMS

Special fruit jams are products similar to marmalades but in which fruit partially keep their shape (whole, halves, etc.). Special fruit jams are obtained by boiling fruit with sugar, with or without pectin addition, with acid addition followed by n concentration by evaporation. Special fruit jams present a pronounced gelification at their cooling and can be considered as fruits included in a pectin-sugar-acid gel. High quality special fruit jams are obtained only from fresh fruit or possibly frozen and from only one fruit species.

Special fruit jams are classified in:

a) non-pasteurized (min. 68% refractometric extract) and

b) pasteurized (min. 65% refractometric extract); minimum acidity, expressed in malic acid, is 0.5%.

Basic Recipe Setting is done taking into account following elements: - maintenance, as much as possible, of fruit shape, because this is specific to these finished products;

- added sugar, in relation to finished product, must be at 60-65%; this high percentage is needed in order to obtain a rapid and strong gelification and to facilitate conservation of fruit shape;

- a satisfactory gelification cannot generally be obtained without pectin addition, at a level of 1-2%, or respectively 10-20% pectic extract (10% refractometric extract);
- partial (about 40%) inversion of added sugar is necessary.

The basic recipe is: 80 kg fruit + 100 kg sugar + 1.6-2.0 kg pectin powder and 1 kg citric or tartric acid; this will yield about 165 kg of special fruit jam with 60% added sugar; water quantity to be evaporated is about 18 kg. If the finished product has to contain 65 % added sugar, the boiling has to be continued up to evaporation of about 30 kg water, the resulting finished product quantity will be about 153 kg.

Technological flow sheet for manufacturing of special fruit jams is as follows:

***Fruit Preparation:*** sorting, washing, peeling and coring (for apples, pears, quinces), or removal of quetches and stones/pits (for plums, peaches, apricots, cherries) or of short tails (for strawberries and wild berries). Pomace fruits are then cut in slices or quarters. Boiling With Sugar is the most important operation in production of special fruit jams and has as objective to evaporate water until gel formation and partial inversion of sugar. Boiling has to be conducted in such a way as to avoid fruit disintegration, but fruits must be well penetrated with sugar. By boiling an equilibrium is reached, by osmosis, between sugar solution and cellular juice. The initial concentration gap between sugar syrup and cellular juice is very high and if the equilibration process is forced, juice comes out of cells and fruit loses its shape and may even disintegrate.

The boiling process accelerates the equilibration, intensity of which increases with temperature and boiling time.

Pectin addition shortens boiling and thus delays the equilibration; for this reason there are different methods for special fruit jams preparation:

- maintaining fruit in sugar, at ambient temperature, over 8-24 hours; fruit to sugar ratio is that indicated in recipe. After this sugar impregnation, fruits and resulting sugar syrup are brought together to boil; toward the end of boiling, pectin, dissolved in warm water, and necessary acid quantity are added.

- preparation of a very concentrated sugar syrup (at least 75%), in which fruit is introduced in order to be boiled.

The rest of the operations are similar to those described in the previous method. Soft fruit (strawberries, wild berries) can be mixed with sugar directly in evaporating open kettles, without added water and then heated gradually up to boiling, which is continued as in previous method.

Boiling is preferably carried out in small open kettles (50-100 kg) in order to avoid too long a boiling and fruit disintegration.

Gelification corresponds generally to the reaching of a concentration of 65% soluble extract, respectively 68% refractometric extract. Practical test for gelification is done as for jellies and marmalades.

Cooling is a technological step strictly necessary in order to avoid fruit rising to the surface in preservation receptacles. Cooling is done in a double bottomed water bath in which water circulates at about 80° C.

Filling of receptacles (jars, boxes, glasses, etc.) is carried out and it is necessary to assure at this stage that the finished product is homogeneous (equal quantities of fruits and gel).

Pasteurization is only applied to special fruit jams with 65 % refractometric extract packed in jars or boxes and is performed at about 100° C for about 20 min.

Gelification is carried out during product cooling and intensifies during storage.

Storage must follow the same conditions as for marmalades.

## NON-GELIFIED SUGAR FRUIT PRESERVES

### Technology of Fruit Jams

Fruit jams are products obtained by boiling of fruits (whole, halves, etc.) with sugar syrup until the reaching a viscous consistency. Jams can be defined as fruits included in a concentrated syrup.

Jams are only prepared from fresh fruits; the usual product range covers the following species: strawberries, cherries, apricots, wild berries, peaches, plums, raisins, quinces, rose petals, etc.; manufacturing of jams from fruit mixes is not an accepted practice. Standards indicate that fruit jams have to conform with following conditions:

- fruit or fruit pieces content 45-55%;
- soluble substances, refractometric degree min. 72%;
- acidity (expressed in malic acid) min. 0.7%.

Basic Recipe Setting is presented below as an indication:

100 kg fruits

150 kg sugar

------

250 kg

35 kg water to be evaporated

------

215 kg jam at about 72% refractometric extract

Fruit Preparation is similar to special fruit jams with the difference that stone / seed removal is mandatory for all species.

Boiling fruits with sugar can be achieved by the same three methods described above for special fruit jams. By boiling a sugar content equilibration is foreseen and the operation must be conducted in such a way that texture, flavour and colour of fruits be preserved. Foam/ scum has to be removed during boiling.

Generally boiling in concentrated sugar syrup (at least 75%) after a previous diffusion during 2-4 hours is in practical use. Boiling must to be carried out in small portions (about 15 kg) in order to avoid fruit disintegration.

For some fruit species, boiling has to be conducted in many phases / steps with "stops" to enable sugar diffusion. At the end of boiling, vanillin at a ratio of 125 g for 100 kg jam may be added for some fruits (white cherries, raisins, etc.). At the same time, it is also possible to add citric or tartric acid in order to avoid the "sugaring" defect.

Cooling of jams, necessary in order to avoid fruit rising to the surface, is carried out as for special fruit jams. Filling of receptacles and Storage for jams are performed in same way as for special fruit jams.

## SPECIAL NON-GELIFIED FRUIT MARMALADES

Special non gelified fruit marmalades are products resulting from fruit without stones or seeds, sieved or squashed, concentrated by boiling, without sugar added and non gelified. Their consistency results from a low water content (about 35%) and a high percentage of insoluble substances (5-10%). Sugar the from fruits acts as a preserving agent.

Plum special non gelified fruit marmalade is the product representative of this category. Other fruit is used very rarely, because they have a reduced sugar content as compared to plums; though there are some countries producing special non gelified fruit marmalade from pears or sweet apples.

For plums, the finished product in this category must contain minimum 55-60% soluble substances (refractometric extract), rising up to 70% for a high quality product.

Basic Recipe Setting for plums is done in relation with sugar content of fruits; when this is higher, the product quality is better and the yield is higher. Thus, if plums with 18% refractometric extract are used, 300-350 kg of fruits are needed for 100 kg finished product.

At the same time, concentration of sugar by boiling, also increases by three fold acidity and astringency of this special plum non gelified marmalade; for this reason it is recommended to use only sweet and completely mature plums.

### Washing is Performed in Usual Conditions

Pre-boiling of fruits can be carried out in water or vapour, preferably with continuous running and has as its objective the softening of tissues.

## De-stoning is Performed in a Pulper

Boiling of the sieved mass is done in double bottom open kettles with a big evaporation surface or in vacuum evaporators.

Boiling in open kettles enables production of a more tasty slightly caramelized, product; boiling in vacuum evaporators has the technological advantages indicated in marmalade production.

At the end of boiling and once of necessary concentration is reached, the product is poured directly into receptacles (drums, etc.) and left to cool in order to form a hard surface layer (crust).

Storage of well closed receptacles is carried out as for marmalade. Special non gelified fruit marmalade can also be prepared from chemically preserved semi-processed fruit products, but the quality is lower than that obtained from fresh fruits. Sometimes dried prunes in a mix with preserved semi-processed products can be used for plum special non gelified marmalade preparation. In some countries plum finished products in this category are sweetened by the addition of maximum 30% sugar, calculated in relation to the finished product.

## TECHNOLOGY OF FRUIT PASTES

These products are obtained in a similar way to marmalades and special non gelified marmalades, but have a lower water content (about 25%). Reduction of water content can be achieved by continuing the boiling of the product or by natural or artificial drying. A typical example of fruit paste without sugar added is the apricot paste - "pistil", etc. which is a concentrated special non gelified fruit marmalade poured in thin layers and sun dried.

An example of fruit paste with sugar added is quince paste which is a marmalade concentrated by evaporation. Sugar content must be 65%; soluble substances content, 7075% refractometric extract and acidity at least 0.5% expressed as malic acid. Packing is done usually in polyethylene sheets and then in boxes or tins; storage conditions are similar to those for marmalade.

## TECHNOLOGY OF FRUIT SYRUPS

Fruit syrups are products obtained by dissolving sugar in juices obtained from direct pressing of fruits. Sugar dissolving can be done at room temperature or by heating.

Syrups have to contain 68% soluble substances (refractometric extract) and minimum 1 g/100 ml malic acid. Up to a maximum 10% of sugar can be replaced by corn syrup. Syrups must be manufactured from the juice of only one fruit species.

Juice Preparation is carried out at room temperature as described above.

Sugar Dissolving In Fruit Juice can be performed by one of the following methods:

(a) `Boiling in open kettles, done by using the basic recipe:

350-400 kg fruit juice;

650-660 kg sugar;

max. 10 kg citric or tartric acid.

The juice is brought to boiling and the sugar is dissolved; the total time has to be as short as possible in order to avoid flavour loss and a too high sugar inversion degree (optimum inversion degree is 40%). Acid is added preferably towards the end of boiling.

During all boiling processes it is necessary to remove foam / scum. In order to avoid caramelization, the syrup has to be cooled rapidly, and this can be carried out in baths with double bottoms through which are circulated water.

One alternative to this method is to boil syrup in closed vessels to avoid flavour losses.

(b) *Boiling in a Vacuum:* The basic recipe is the same as above. Sugar and fruit juice are mixed previously in a pre-heating kettle and then transported to vacuum equipment.

Boiling is performed at 50° C and at the end the temperature is raised slowly up to 65-70° C. The syrup can be cooled directly in vacuum equipment by closing the steam inlet and by increasing the vacuum. In this boiling method it is possible to incorporate a flavour recuperation device.

(c) A continuous process for syrup preparation can be carried out by dissolving components with heat while passing them through a horizontal cylinder with a screw inside.

In the methods where sugar is dissolved by heat, it is also possible to use chemically preserved juices.

In this case it is necessary to first perform the desulphitation of juices preserved with SO2. This can be performed by boiling juice with optional water addition (and before any sugar addition). High quality syrups are obtained however from fresh juices.

(d) Sugar can also be dissolved at room temperature by using continuous flow percolators. These are similar to those used for salt solution preparation in vegetable canning processes. The juice goes over a sugar layer and is concentrated progressively until saturation (about 65 %). The syrup is then passed through a filtration section in the bottom of the percolator.

Syrup Filtration is needed in order to clarify crystals; the filtration of syrup is done in warm conditions through cloth. Filling of syrup in bottles is done in aseptic conditions as much as possible in order to avoid syrup infection with osmophile yeasts. Syrup preservation is assured by the high sugar content with respect to a low water activity unclear. Storage takes place in well ventilated storage rooms; avoiding sunlight at 10-15° C. The usual product range is: strawberries, cherries, wild berries, citrus fruits.

## TECHNOLOGY OF "FRUIT IN SYRUP" PRODUCTS

This type of product is represented by fruit (whole, halves or pieces) covered by a sugar solution and preserved by pasteurization. In these products sugar does not have a preservation effect but only a sweetening role.

Technological general flow-sheet covers the following steps:

Sorting is necessary in order to choose mature fruit, whole, undamaged and with a specific texture. Fruit of good quality but with a texture not compatible with this type of finished product are used for the production of semi-processed products, marmalades, etc. This step is done manually on a sorting belt.

Washing is performed in equipment with fan and sprays. Cleaning covers removal of leaves, etc., skin removal by one of the described methods for some fruits and coring / pitting for others. Cutting is applied only to pomace fruits and is performed preferably by mechanical means.

Preliminary Heat Treatment depends very much upon the fruit types and can be as different as a light blanching up to a real boiling; this step is aimed toward softening of tissues for hard fruits, elimination of waxy pruin layer (plums, etc.) or enzyme inactivation for pomace fruits.

This treatment has to be reduced to the minimum necessary in order to avoid sugar losses. Some fruit (for example apricots, black cherries, grapes, etc.) does not undergo preliminary heat treatment.

Cooling is carried out in water (as cold as possible) and should not be too long to avoid soluble substances loss.

Receptacle Filling Fruit is introduced manually or sometimes mechanically (filling tables, etc.) in receptacles and then sugar syrup is added.

Sugar Syrup Preparation is performed by dissolving crystal sugar in hot water (90100° C). After the sugar has been dissolved, the syrup is boiled briefly; removal of impurities and coagulated substances is then performed by foaming/scumming. An addition of about 0.3% citric acid helps syrup clarification, followed by filtration through cloth.

Preheating/Emptying of open receptacles has as its objective to eliminate air from fruit tissues and this step is largely replaced today by receptacle closing under vacuum. Preheating may be done in steam or in water in such a way that syrup temperature reaches 8090° C and is maintained for 10-15 min.

Hermetic Sealing is performed according to the type of receptacle and of cover/cap.

Pasteurization is carried out at 100° C and may be performed in continuous installations or in normal water baths. In some countries fruits in syrup are pasteurized at 80-90° C; this contributes to a greater flavour retention.

Cooling should be intensive in order to avoid discolorations and colour modifications and to reduce corrosion in metal cans.

Refractometric extract is to be measured 30 days after manufacturing. In order to establish exact concentration of sugar syrup to be added in each individual case it is possible to use the following formula:

$C = 1.5 \times [ (2 \times E) - F ]$

C = added syrup concentration;

E = refractometric extract needed in finished products, according to Table.

F = fruit refractometric extract.

Hot syrup is put into receptacles with a minimum temperature of 80° C in order to enable a corresponding vacuum after cooling.

It is possible to add in the syrup 0.3-0.5% citric or tartaric acid for fruit with a low acidity. Adding ascorbic acid in proportion of 0.8% assures colour maintenance and taste improvement based on its reducing action. Avoidance of excessively soft texture in finished products may be achieved for soft fruit (strawberries, apricots, etc.) by dipping fruit in a solution of calcium chloride (5% $CaCl_2$ in water).

Finished product defects and production "accidents" for "Fruit in syrup" and the means of preventing them

- Change of colour in red fruit (red cherries, plums, etc.) to violet; to avoid this it is necessary to use glass jars or varnished/lacquered tinplate;
- Colour change to red for pears and quinces (pink shade) could arise in products which are over-pasteurized and insufficiently cooled;
- Change of colour of whole peaches peeled by chemical means is due to an incomplete inactivation of oxidative

enzymes; to avoid this it is necessary to boil fruit after chemical peeling followed by washing with 0.5% citric or tartric acid solutions or by adding ascorbic acid in syrup;

- Softening of strawberries tissues can be prevented by a pre-treatment with warm syrup and also by addition of calcium salt.

Some technical data for processing are seen in Table.

***Technical Data for "Fruits in Syrup"***

| Type of fruit/product | | Minimum fruit | Soluble substances (in syrup), RE** at 20°C, |
|---|---|---|---|
| name | Size/shape | content, % | minimum |
| Apricots | Halves | 55 | 25 |
| Strawberries | Whole | 47 | 25 |
| Cherries | Whole | 53 | 22 |
| Quinces | Slices, 30 mm | 50 | 26 |
| Apples | Halves or quarters | 47 | 25 |
| Pears | Halves or quarters | 57 | 22 |
| Plums | Whole | 45 | 25 |
| Melon | Pieces | 55 | 26 |
| Wax cherries | Whole | 53 | 27 |
| Wild cherries | Whole | 47 | 27 |
| Pineapple | Slices | 55 | 26 |
| Mangos | Pieces | 50 | 25 |
| Papaya | Pieces | 55 | 27 |
| Fruit mixes | According to the conditions indicated above, for each fruit | 50 | 25 |

** RE = Refract metric extract

## Fruit Juice Technologies

Fruit juices are products for direct consumption and are obtained by the extraction of cellular juice from fruit, this operation can be done by pressing or by diffusion.

For the purpose of this document, the technology of fruit juice processing will cover two finished product categories:

- juices without pulp ("clarified" or "not clarified");
- juices with pulp ("nectars").

We will also define as :

a. "natural juices" products obtained from one fruit; and

b. "mixed juices" products obtained from the mix of two or three juices from different fruit species or by adding sugar.

Juices obtained by removal of a major part of their water content by vacuum evaporation or fractional freezing will be defined as "concentrated juices".

## Technological Steps for Processing of Fruit Juices without Pulp

Fruit juices must be prepared from sound, mature fruit only. Soft fruit varieties such as grapes, tomatoes and peaches should only be transported in clean boxes which are free from mould and bits of rotten fruit.

***Washing:*** Fruit must be thoroughly washed. Generally, fruit will be submitted to a pre-washing before sorting and a washing step just after sorting.

***Sorting:*** Removal of partially or completely decayed fruit is the most important operation in the preparation of fruit for production of first quality fruit juices; sorting is carried out on moving inspection belts or sorting tables.

Crushing/Grinding/Disintegration Step is applied in different ways and depends on fruit types:

Crushing for grapes and berries;

Grinding for apples, pears;

Disintegration for tomatoes, peaches, mangoes, apricots.

This processing step will need specific equipment which differs from one type of operation to another.

Enzyme Treatment of crushed fruit mass is applied to some fruits by adding 2-8% pectolitic enzymes at about 50° C for 30 minutes.

This optional step has the following advantages: extraction yield will be improved, the juice colour is better fixed and finished product taste is improved.

However, for fruit which is naturally rich in pectic substances, this treatment makes the resulting "exhausted" material useless for industrial pectin production.

Heating of crushed fruit mass before juice extraction is an optional step used for some fruit in order to facilitate pressing and colour fixing; at same time, protein coagulation takes place.

Pressing to extract juice.

Diffusion is an alternative step for juice extraction and can be carried out discontinuously or in batteries at water temperature of about 80-85°C.

Juice Clarifying can be performed by centrifugation or by enzyme treatment. Centrifugation achieves a separation of particles in suspension in the juice and can be considered as a pre-clarifying step. This operation is carried out in centrifugal separators with a speed of 6000 to 6500 RPM.

Enzyme clarifying is based on pectic substance hydrolysis; this will decrease the juices' viscosity and facilitate their filtration. The treatment is the addition of pectolitic enzyme preparations in a quantity of 0.5 to 2 g/l and will last 2 to 6 hours at room temperature, or less than 2 hours at 50° C, a temperature that must not be exceeded.

The control of this operation is done by checking the decrease in juice viscosity. Sometimes, the enzyme clarifying is completed with the step called "sticking" by the addition of 5-8 g/hl of food grade gelatine which generates a flocculation of particles in suspension by the action of tannins.

Filtration of clarified juice can be carried out with kieselgur and bentonite as filtration additive in press-filters (equipment).

De-tartarisation is applied only to raisin juice and is aimed to eliminate potassium bitartrate from solution. This step can be performed by the addition of 1% calcium lactate or 0.4% calcium carbonate.

Pasteurization of juice can be done for temporary preservation (pre-pasteurization) and in this case this operation is carried out with continuous equipment (heat exchangers, etc.); warm juice is stored in drums or large size receptacles

(20-30 kg). Pasteurization conditions are at 75°C in continuous stream.

Pasteurization of bottled juice is then carried out just before delivery to the market; this is performed in water baths at 75° C until the point where the juice reaches 68° C. In cases when the final pasteurization is done without pre-pasteurization and temporary storage, modern methods use a rapid pasteurization followed by aseptic filling in receptacles.

Rapid pasteurization conditions are as follows: temperature about 80° C, over 10-60 sec., followed by cooling; all operations are carried out in continuous stream.

Preservation under $CO_2$ pressure may be done at a concentration of 1.5% $CO_2$ under a pressure of 7 kg/cm$^2$. At the distribution step, proceed at $CO_2$ decompression and the juice is then submitted to a sterilising filtration and aseptic filling in receptacles.

Preservation by freezing is carried out at about -30° C, after a preliminary de-aeration; storage is at -15 to -20° C.

Production of concentrated juices by evaporation is performed under vacuum (less than 100 mm Hg residual pressure) up to a concentration of 65-70% total sugar which assures preservation without further pasteurization. Modern evaporation installations recover flavours from juices which are then reincorporated in concentrated juices.

Additional operations for juice manufacturing are the vacuum de-aeration and mixing with other fruit juices or with sugar. For the production of non clarified juices the centrifugation is the only specific step, enzyme clarifying and subsequent filtration being eliminated.

The optimum sugar/acid ratio for the majority of fruit, mainly for pomaces, is 10/1 to 15/1.

Fruit which is rich in carotenoids (apricots, peaches, etc.) is only processed as juices with pulp ("nectars").

Technological flow-sheet for fruit juices with pulp ("Nectars")

This process is divided at industrial scale in two categories of operations:

a. processing for obtaining juices;

b. juice conditioning for preservation.

(a) Operations in the first category are differ according to the type of fruit which to be processed.

Pomaces (apples, pears) are washed and sorted and then crushed in a colloid mill; fruit puree is then passed through a screw type heating equipment where direct steam is used as a source of heat. Warm fruit mass is treated in a pulper with a 2 mm screen and then through an extractor similar with the equipment used for tomato juice.

Stone fruits (apricots, peaches, cherries, etc.) after washing and sorting are submitted to steam in a continuous heater, then the warm fruit mass is passed through a pulper and then an extractor (as mentioned above). Berries (strawberry, wild berries, etc.) are washed, sorted and then crushed, preheated and then introduced in extractor.

In order to avoid browning and undesirable taste modifications it is usual to add about 0.05% ascorbic acid.

(b) Second category type of operations are similar for all fruit species. Partial elimination of cellulose is achieved with a continuous centrifugal separator; the resulting juice is then processed in order to adjust sugar and acid content for viscosity.

Sugar (about 8-10%) is added as a syrup (in water or in the juice of same fruit obtained by pressure). Acidity is adjusted with citric or tartaric acid. The adjusted juice is then deaerated under vacuum at about 40° C. This step aims at avoiding oxidative reactions and vitamin C loss reduction.

An important subsequent step is an intensive homogenisation (under pressure at 150-180 A) in order to obtain particles with dimensions below 100. The homogenised juice obtained is then continuously pasteurized in plate heat exchanger equipment at a temperature of about 130° C, cooled down to about 90° C and aseptically packed in receptacles.

The principal characteristics of fruit "nectars" are uniformity and stability of the content provided by the advanced disintegration of fruits. Stability can be obtained by increasing product viscosity by adding pectin for fruit which is deficient in this component. In order to avoid "separation", intensive homogenisation is carried out as described above.

Fruit "nectars" contain all the important components of the original fruit and to a large extent maintain their taste and flavour. The sugar/acidity (as citric acid) ratio is to a large extent determined by the type of fruit and the correction applied; for example, this ratio is 30 for apricots, 40 for peaches, 160 for pears, etc.

A general technological flow-sheet for fruit juice processing is presented below.

## BANANA AND PLANTAIN PROCESSING TECHNOLOGIES

### Traditional Processing

1 *Products:* Uses and Dietary Significance

Most of the world's bananas are eaten either raw, in the ripe state, or as a cooked vegetable, and only a very small proportion are processed in order to obtain a storable product. This is true both at a traditional village level with both dessert and cooking bananas and when considering the international trade in dessert bananas.

In general, preserved products do not contribute significantly to the diet; however, in some localised areas the products are important in periods when food are scarce.

Probably the most widespread and important product is flour preparation from unripe banana and plantains by sun-drying. In Uganda, dried slices known as "mutere" are prepared for storage from green bananas, the dried slices being either used directly for cooking or after grinding into a flour. "Mutere" is used chiefly as a famine reserve and does not feature largely in the diet under normal conditions.

In Gabon, plantains are sometimes made into dried slices which can be stored and used on long journeys, and plantains are used in Cameroon to prepare dried pieces which are stored and ground as needed into flour for use in cooking a paste known as "fufu". Dried green banana slices are also used in parts of South and Central America and West Indies for preparing flour.

The other nutritionally important product is beer which is a major product in Uganda, Rwanda and Burundi where green banana utilisation is particularly high.

2. Preservation Methods and Processes.

***Drying.*** Both ripe and unripe bananas and plantains are normally peeled and sliced before drying, although banana figs are sometimes prepared from whole ripe fruit. Sun drying is the most widespread technique where the climate is suitable but drying in ovens or over fires is also practiced. In west Africa, plantains are often soaked and sometimes parboiled before drying.

The slices of unripe fruit are normally spread out on bamboo frameworks; or a cemented area; or on a mat; or on a swept-bare patch of earth; or on a roof; or sometimes on stones outcrops or sheets of corrugated iron.

Oven-drying of ripe bananas is practiced in Polynesia as a mean of preserving the fruits, which are then wrapped in leaves and bound tightly to store until needed. In East Africa a method has been reported that involves drying the peeled bananas on a frame placed over a fire for 24 hr before drying in the sun, to accelerate the process.

3. Product stability and storage problems

There is little experimental data on the storage life of the traditionally made banana and plantain products.

4. Potential for scaling up of traditional processes to industrial level

Many banana products are now produced on an industrial scale, including the traditional banana figs and flour, and the processing techniques are described below. One of the main problems encountered has been the susceptibility of banana products to flavour deterioration and discoloration and in the past many products reaching the market have been of poor quality.

A great deal of research has been directed to overcoming these problems, although however good the resultant products are they cannot compare in flavour and other characteristics with the fresh banana fruit. Indeed, an important constraint on the large-scale development of banana processing is the lack of demand for banana products since the fresh fruit is available throughout the year in most parts of the tropical world.

The production of beer from banana and plantains has not been scaled up to an industrial level, and while an important product in localised areas of tropical Africa, the market is rapidly declining in favour of European-type brews produced locally.

## INDUSTRIAL PROCESSING

***Products and Uses:*** The main commercial products made from bananas are canned or frozen puree, dried figs, banana powder, flour, flakes, chips (crisps), canned slices and jams.

Banana products can be divided roughly into two types - those for direct consumption, such as figs, and those for use in food manufacturing industry, for example purees and powder.

Banana figs, or fingers as they are sometimes known, are usually whole, peeled fruit carefully dried so as to retain their shape, although sometimes the fruit is sliced or halved to facilitate drying. Banana and plantain chips (crisps) are thinly sliced pieces of fruit fried in oil and eaten as a snack like potato chips (crisps).

The main use of canned slices is in tropical fruit salads. Banana flakes are used as a flavouring or in breakfast cereals. Banana puree find use mainly in the production of baby foods. Banana flour is said to be highly digestible and is used in baby and invalid foods, but can also be used in the preparation of bread and beverages.

Banana powder is used chiefly in the baking industry for the preparation and fillings for cakes and biscuits and is also used for invalid and baby foods.

***Processing Technology:*** In general, to obtain a good-quality product from ripe-bananas the fruit is harvested green and ripened artificially under controlled conditions at the processing factory. After ripening, the banana hands are washed to remove dirt and any spray residues, and peeled. Peeling is almost always done by hand using stainless steel knives, although a mechanical peeler for ripe bananas has been developed, capable of peeling 450 Kg of fruit per hour (Banana Bulletin, 1974).

The peeling of unripe bananas and plantains is facilitated by immersing the fruit in hot water. For example, immersion in water at 70-75 ° C for 5 min. has been suggested as an aid for peeling green bananas for flour production, while the peeling of green bananas for freezing has been facilitated by immersion in water at 93° C for 30 min.

***Banana Figs:*** Fully ripe fruits with a sugar content of about 19.5% are used and are treated with sulphurous acid after peeling, then dried as soon as possible afterwards. Various drying systems have been described using temperatures between 50 and 82° C for 10 to 24 hr to give a moisture content ranging from 8 to 18% and a yield of dried figs of 12 to 17% of the fresh banana on the stem.

One factory in Australia uses a solar heat collector on the roof to augment the heat used for drying bananas. Bananas can also be dried by osmotic dehydration, using a technique which involves drawing water from 1/4-in. thick banana by placing them in a sugar solution of 67 to 70 deg. Brix for 8 to 10 hr. followed by vacuum-drying at 65 to 70° C, at a vacuum of 10 mm Hg for 5 hr. The moisture content of the final products is 2.5% or less, much lower than that achieved by other methods.

***Banana Puree:*** Banana puree is obtained by pulping peeled, ripe bananas and then preserving the pulp by one of three methods: canning aseptically, acidification followed by normal canning, or quick-freezing.

The bulk of the world's puree is processed by the aseptic canning technique. Peeled, ripe fruits are conveyed to a pump which forces them through a plate with 1/4-in. holes, then onto a homogeniser, followed by a centrifugal de-aerator, and into a receiving tank with 29in. vacuum, where the removal of air helps prevent discoloration by oxidation. The puree is then passed through a series of scraped surface heat exchangers where it is sterilised by steam, partially cooled, and finally brought to filling temperature. The sterilised puree is then packed aseptically into steam-sterilised cans which are closed in a steam atmosphere.

***Banana Slices:*** Several methods for canning of banana slices in syrup are used. Best-quality slices are obtained from fruit at an early stage of ripeness. The slices are processed in a syrup of 25 deg. Brix with pH about 4.2, and in some processes calcium chloride (0.2%) or calcium lactate (0.5%) are added as firming agents.

A method for producing an intermediate-moisture banana product for sale in flexible laminate pouches has been developed. Banana slices are blanched and equilibrated in a solution containing glycerol (42.5%), sucrose (14.85%), potassium sorbate (0.45%), and potassium metabisulphite (0.2%) at 90 deg. C for 3 min. to give a moisture content of 30.2%.

***Banana Powder:*** In the manufacture of banana powder, fully ripe banana pulp is converted into a paste by passing through a chopper followed by a colloid mill. A 1 or 2 % sodium metabisulphite solution is added to improve the colour of the final product. Spray- or drum-drying may be used, the latter being favoured as all the solids are recovered.

A typical spray dryer can produce 70 kg powder per hour to give yields of 8 to 11% of the fresh fruit, while drum-drying gives a final yield of about 13% of the fresh fruit. In the latter method the moisture content is reduced to 8 to 12 % and then further decreased to 2 % by drying in a tunnel or cabinet dryer at 60° C.

***Banana Flour:*** Production of flour has been carried out by peeling and slicing green fruit, exposure to sulphur dioxide gas, then drying in a counter-current tunnel dryer for 7 to 8 hr. with an inlet temperature of 75° C and outlet temperature of 45° C, to a moisture content of 8%, and finally milling.

***Banana Chips (crisps):*** Typically, unripe peeled bananas are thinly sliced, immersed in a sodium or potassium metabisulphite solution, fried in hydrogenated oil at 180 to 200° C, and dusted with salt and an antioxidant.

Alternatively, slices may be dried before frying and the antioxidant and salt added with the oil. Similar processes for producing plantain chips have been developed.

***Banana Beverages:*** In a typical process, peeled ripe fruit is cut into pieces, blanched for 2 min. in steam, pulped and pectolytic enzyme added at a concentration of 2 g enzyme per 1 kg pulp, then held at 60 to 65° C and 2.7 to 5.5 pH for 30 min.

In a simpler method, lime is used to eliminate the pectin. Calcium oxide (0.5%) is added to the pulp and after standing for 15 min. this is neutralised giving a yield of up to 88% of a clear, attractive juice. In another process banana pulp is acidified, and steam-blanched in a 28-in Hg vacuum which ensures disintegration and enzyme inactivation. The pulp is then conveyed to a screw press, the resulting puree diluted in the ratio 1:3 with water, and the pH adjusted by further addition of citric acid to 4.2 to 4.3, which yields an attractive drink when this is centrifuged and sweetened.

***Jam:*** A small amount of jam is made commercially by boiling equal quantities of fruit and sugar together with water and lemon juice, lime juice or citric acid, until setting point is reached.

## Product Stability and Spoilage Problems

All dried banana products are very hydroscopic and susceptible to flavour deterioration and discoloration, but this can be overcome to some extent by storing in moisture-proof containers and sulphiting the fruit before drying to inactivate the oxidases. The dried products are also liable to attack by insects and moulds if not stored in dry conditions, although disinfestation after drying by heating for 1 hr to 80° C or by fumigation with methyl bromide ensures protection against attack. Banana powder is said to be stored for up to a year commercially and flakes have been stored in vacuum-sealed cans with no deterioration in moisture, colour or flavour for 12 months.

Banana chips tend to have a poor storage life and to become soft and rancid. However, chips treated with an antioxidant have been stored satisfactorily at room temperature in hermetically sealed containers up to 6 months with no development of rancidity.

## Quality Control Methods

In general a good quality product is obtained if fruit is harvested at the correct stage of maturity and, where appropriate, ripened under controlled conditions. For example, in the case of banana figs, the fruit should be fully mature (sugar content of 19.5% or above) or the final product is liable to be tough and lacking in flavour. However, if over-ripe fruit is used, the figs tend to be sticky and dark in colour, so the fruit must be fully yellow but still firm.

For banana flour, which is prepared from unripe bananas, the fruit is harvested at three-quarters the full-ripe stage and is processed within 24 hr. prior to the onset of ripening. If less mature fruit is used, the flour tastes slightly astringent and bitter due to the tannin content. Bananas harvested between 85 and 95 days after the emergence of the inflorescence, with a pulp-to-peel ratio of about 1.7, were considered to be most suitable for the deep-fat frying.

Other criteria suggested for assessing maturity were beta-carotene and reducing sugar content, both of which increase with increasing maturity and pH which decreases as the fruit ripens, and these should be, respectively, about 2000 μg/100 g, less than 1.5% and 5.8 or above. Browning was found to occur if the sugar content was higher than 1.5%. The determination of crude fat in processed chips is also considered to be a necessary quality control measure. It is important to remove all impurities prior to processing of products, and this is done by washing to remove dirt and spray residues and control on the processing line so that substandard fruit can be removed.

## Preparation Methods for Fresh Bananas and Plantains

The main ways of preparing fresh bananas for consumption are boiling or steaming, roasting or baking and frying. Boiling followed by pounding into "fufu" is also widely adopted in certain areas of the tropics.

***Boiling or Steaming:*** Plantains and bananas are often prepared simply by boiling in water, either in their peel or after peeling, and either ripe or unripe; if unripe, the fruit is scraped thoroughly after peeling to remove all traces of fibrous material. The boiled fruit is eaten alone or more usually accompanied by a sauce. This preparation technique is widely used in West Africa.

***Roasting or Baking:*** Unpeeled or peeled fruit, either ripe or unripe, is roasted simply by placing in the ashes of a fire or in an oven. This method is widely used in West Africa, East Africa and the South Pacific islands. For example, ripe plantains are placed unpeeled in an oven and when partly brown and tender, removed and peeled, then replaced in the oven and roasted evenly.

***Frying:*** Ripe or unripe plantains or bananas are often peeled, sliced and cooked in oil, particularly in West Africa and in parts of South America and the West Indies. Similar products are also made in East Africa. Typically, ripe plantains are peeled, cut into slices or split lengthways, and fried in palm oil or with groundnut oil, the pieces being served either hot with a sauce or with fried eggs, or cold as a snack.

***Pounding:*** Pounding is a process, used particularly in West Africa, for preparing most perishable staple food crops including plantains, cassava, yams and cocoyams to obtain a paste or dough known as "fufu" (also spelled "foofoo", "foutou", "foufou"). The plantains are peeled or boiled and peeled after boiling and pounded in a wooden mortar, the resulting paste normally being eaten with soup or a spiced sauce of meat and vegetables, but sometimes after wrapping in leaves and steaming.

## MANGO AND GUAVA PROCESSING TECHNOLOGIES

Mangoes are processed at two stages of maturity. Green fruit is used to make chutney, pickles, curries and dehydrated products. The green fruit should be freshly picked from the tree. Fruit that is bruised, damaged, or that has prematurely fallen to the ground should not be used. Ripe mangoes are processed as canned and frozen slices, puree, juices, nectar and various dried products. Mangoes are processed into many other products for home use and by cottage industry.

The mango processing presents many problems as far as industrialization and market expansion is concerned. The trees are alternate bearing and the fruit has a short storage life; these factors make it difficult to process the crop in a continuous and regular way. The large number of varieties with their various attributes and deficiencies affects the quality and uniformity of processed products.

The lack of simple, reliable methods for determining the stage of maturity of varieties for processing also affects the quality of the finished products. Many of the processed products require peeled or peeled and sliced fruit. The lack of mechanised equipment for the peeling of ripe mangoes is a serious bottleneck for increasing the production of these products.

A significant problem in developing mechanised equipment is the large number of varieties available and their different sizes and shapes. The cost of processed mango products is also too expensive for the general population in the areas where most mangoes are grown. There is, however, a considerable export potential to developed countries but in these countries the processed mango products must compete with established processed fruits of high quality and relatively low cost.

## Green Mango Processing

***Pickles:*** The optimum stage of maturity should be determined for each variety used to make pickles.

There are two classifications of pickles - salt pickles and oil pickles. They are processed from whole and sliced fruit with and without stones. Salt is used in most pickles.

The many kinds of pickles vary mainly in the proportions and kinds of spices used in their preparation. One basic recipe for the study of the preparation and storage of pickles in oil is as follows:

| | | | |
|---|---|---|---|
| Mango pieces | 250 g | Tumeric powder | 2 to 4 g |
| Salt | 60 g | Fenugreek seeds | 2 to 4 g |
| Mustard powder | 30 g | Bengal gram seeds | 2 to 4 g |
| Chili powder | 20 g | Gingelly oil | 20 to 30 g |

The ingredients are mixed together and filled into wide-mouthed bottles of 0.5 kg capacity. Three days later the contents are thoroughly mixed and refilled into the bottles. Extra oil is added to form a 1-2 cm layer over the pickles.

***Chutney:*** The product is prepared from peeled, sliced or grated unripe or semi-ripe fruit by cooking the shredded fruit with salt over medium heat for 5 to 7 minutes, mixed and then sugar, spices and vinegar are added. Cook over moderate heat until the product resembles a thick puree, add remaining ingredients and simmer another 5 min. Cool and preserve in sterilised jars. Spices usually include cumin seeds, ground cloves, cinnamon, chili powder, ginger and nutmeg. Other ingredients such as dried fruits, onions, garlic and nuts may be added.

***Drying/dehydration:*** Immature fruit is peeled and sliced for sun-drying. The dried mango slices can be powdered to make a product called amchoo. The use of blanching, sulphuring and mechanical dehydration gives a product with better colour, nutrition, storability and fewer microbiological problems.

**Ripe Mango Processing**

***Puree:*** Mangoes are processed into puree for re-manufacturing into products such as nectar, juice, squash, jam, jelly and dehydrated products. The puree can be preserved by chemical means, or frozen, or canned and stored in barrels. This allows a supply of raw materials during the remainder of the year when fresh mangoes are not available.

It also provides a more economical means of storage compared with the cost of storing the finished products, except for those which are dehydrated, and provides for more orderly processing during peak availability of fresh mangoes.

Mangoes can be processed into puree from whole or peeled fruit. Because of the time and cost of peeling, this step is best avoided but with some varieties it may be necessary to avoid off-flavours which may be present in the skin. The most common way of removing the skin is hand-peeling with knives but this is time-consuming and expensive. Steam and lyepeeling have been accomplished for some varieties.

Several methods have been devised to remove the pulp from the fresh ripe mangoes without hand-peeling. A simplified method is as follows: the whole mangoes were exposed to atmospheric steam for 2 to 2 1/2 min in a loosely covered chamber, then transferred to a stainless steel tank.

The steam-softened skins allowed the fruit to be pulped by a power stirrer fitted with a saw-toothed propeller blade mounted 12.7 to 15.2 cm below a regular propeller blade. The pulp is removed from the seeds by a continuous centrifuge designed for use in passion fruit extraction. The pulp material is then passed through a paddle pulper fitted with a 0.084 cm screen to remove fibre and small pieces of pulp.

Mango puree can be frozen, canned or stored in barrels for later processing. In all these cases, heating is necessary to preserve the quality of the mango puree. In one process, puree is pumped through a plate heat exchanger and heated to 90°C for 1 min and cooled to 35° C before being filled into 30 lb tins with polyethylene liners and frozen at -23.50 C.

In an other process, pulp is acidified to pH 3.5, pasteurized at 90°C, and hot-filled into 6 kg high-density bulk polyethylene containers that have been previously sterilised with boiling water. The containers are then sealed and cooled in water. This makes it possible to avoid the high cost of cans.

Wooden barrels may be used to store mango pulp in the manufacture of jams and squashes. The pulp is acidified with 0.5 to 1.0% citric acid, heated to boiling, cooled, and $SO_2$ is added at a level of 1000 to 1500 ppm in the pulp. The pulp is then filled into barrels for future use.

***Slices:*** Mango slices can be preserved by canning or freezing, and recent studies have shown the feasibility of pasteurized-refrigerated and dehydro-canned slices. The quality of the processed product in all of these procedures will be dependent upon selection of a suitable variety along with good processing procedures. Thermal process canning of mango slices in syrup is the most widely used preservation method.

***Beverages:*** The commercial beverages are juice, nectar and squash. Mango nectar and juice contain mango puree, sugar, water and citric acid in various proportions depending on local taste, government standards of identity, pH control, and fruit composition of the variety used. Mango squash in addition to the above may contain SO2 or sodium benzoate as a preservative. Other food grade additives such as ascorbic acid, food colouring, or thickeners may be used in mango beverages.A short description of finished products found in literature is as follows:

- *Mango Juice:* prepared by mixing equal quantities of pulp (puree) and water together and adjusting the total soluble solids (TSS) and acidity to taste (12 to 15% TSS and 0.4 to 0.5% acidity as citric acid);
- Mango nectar containing 25% puree can be prepared using the following procedure.

| | Brix of purée | | |
|---|---|---|---|
| Nectar components | 15° | 17° | 20° |
| Purée | 100 | 100 | 100 |
| Sugar | 45 | 43 | 40 |
| Water | 255 | 257 | 260 |

Commercial processing conditions may require the use of a preservative.

The pH is adjusted to approximately 3.5 by adding citric acid as a 50% solution.

The time of heat processing will vary with filling temperature, can size and viscosity of the juice or nectar.

Mango squash may be prepared according to flow-sheet described below; the finished product may contain 25% juice, 45% TSS and 1.2 to 1.5% acidity and may be preserved with sulphur dioxide (350ppm) or sodium benzoate (1000 ppm) in glass bottles.

Mango squash simplified flow-sheet.

### *Ingredients*

| Mango pulp | 900 | 900 |
|---|---|---|
| Sugar | 900 | 1100 |
| Citric acid | 18 | 15 |
| Water | 900 | 900 |

Mangoes are washed, stored, peeled with stainless steel knives. The pulp is prepared by using a pulper with fine sieve (0.025-in); Sugar is mixed with water and citric acid = syrup; The pulp is added to the syrup and mixed well; The mixture is strained trough cloth; The squash is heated at 85° C and bottles are filled and closed. For additional heat treatment bottles may need to be maintained at a product temperature of 80°C for 30 minutes if the product is to be processed without preservatives. The bottles are then left to cool in water and stored at room temperature.

Two negative points must be avoided: presence of air bubbles (which is a source of quick deterioration) and separation of squash solids (giving an undesirable appearance). The means to avoid these two phenomena are described in the fruit juices section. A type of "squash type" beverage may also be manufactured with '/a pulp + '/a water + i/a sugar and pH adjusted to 3.7 by addition of citric acid. Using different sieve sizes affects the quality and reduces air bubbles to a certain extent but homogenisation and de-aeration of puree or squash seem to be important in order to avoid separation and air bubbles. The squash quality is evaluated on the basis of the following characteristics: pH, titrable acidity, soluble solids, ascorbic acid (by 2,6 dichlorophenol indophenol method), specific gravity.

***Dried/dehydrated:*** Ripe mangoes are dried in the form of pieces, powders, and flakes. Drying procedures such as sun-drying, tunnel dehydration, vacuum-drying, osmotic dehydration may be used. Packaged and stored properly, dried mango products are stable and nutritious.

One described process involves as pre-treatment dipping mango slices for 18 hr (ratio 1:1) in a solution containing 40°

Brix sugar, 3000 ppm $SO_2$, 0.2% ascorbic acid and 1% citric acid; this method is described as producing the best dehydrated product. Drying is described using an electric cabinet through flow dryer operated at 60° C. The product showed no browning after 1 year of storage.

Drum-drying of mango puree is described as an efficient, economical process for producing dried mango powder and flakes. Its major drawback is that the severity of heat preprocessing can produce undesirable cooked flavours and aromas in the dried product. The drum-dried products are also extremely hydroscopic and the use of in-package desiccant is recommended during storage.

***Canning:*** This preservation technology is described in various technological flow-sheets in this bulletin. Mango bar or "fruit leather" is presented in various flow-sheets.

## GUAVA PROCESSING TECHNOLOGIES

***Guava Puree:*** Guava puree is used in the manufacture of guava nectar, various juice drink blends and in the preparation of guava jam. The washed sound fruit is first passed through a chopper or slicer to break up the fruit and this material is fed into a pulper. The pulper will remove the seeds and fibrous pieces of tissue and force the reminder of the product through a perforated stainless steel screen. The holes in the screen should be between 0.033 and 0.045 in. The machine should be fed at a constant rate to ensure efficient operation.

The pureed material coming from the pulper is next passed through a finisher. The finisher is equipped with a screen containing holes of approximately 0.020 in. The finisher will remove the stone cells from the fruit and provide the optimum consistency to the product. Perhaps the best way to preserve the guava puree is by freezing and the material passing through the finisher can be packaged and frozen with no further treatment. It is not necessary to heat the product to inactivate enzymes or for other purposes. The material can be frozen in a number of types of cartons and cans; however, a fibre box with a plastic bag inside is commonly used and is probably the less expensive.

It is also possible to can and heat process the guava puree and this can be accomplished by heating the puree to 195° F in an open double bottom kettle, filling into cans, closing the cans, inverting the cans for a few seconds, followed by cooling. Cans should be cooled rapidly to approximately 100° F before they are cased and stacked into warehouses.

***Guava Juice and Concentrate:*** Guava juice can be used in the manufacture of a clear guava jelly or in various drinks. A clear juice may be prepared from guava puree that is depectinised enzymatically. About 0.1% pectin-degrading enzyme is mixed into the puree at room temperature; heating of the product at approximately 120° F will greatly speed the action of the enzyme. After 1 hr. clear juice is separated from the red pulp by centrifuging or by pressing in a hydraulic juice press. A batch-type or continuous-flow centrifuge can be used on the depectinised puree with no further treatment.

The clear juice after centrifuge or after press (and subsequent filtration) can be preserved by freezing or by pasteurization in hermetically sealed cans. For shipment to overseas markets it may be advantageous to concentrate either the puree or the juice.

## RECENT TRENDS IN FRUIT AND VEGETABLE PROCESSING

***New Products:*** The number and variety of fruit and vegetable products available to the consumer has increased substantially in recent years. The fruit and vegetable industry has undoubtedly benefited from the increased recognition and emphasis on the importance of these products in a healthy diet. Traditional processing and preservation technologies such as heating, freezing and drying together with the more recent commercial introduction of chilling continue to provide the consumer with increased choice. This has been achieved by new heating (*e.g.* UHT, microwave, ohmic) and freezing (*e.g.* cryogenic) techniques combined with new packaging materials and technologies (*e.g.* aseptic, modified atmosphere packaging).

The overall trend in new fruit and vegetable products is "added value", thus providing increased convenience to the consumer by having much greater variety of ready prepared fruit and vegetable products. These may comprise complete meals or individual components. The suitability of products and packages for microwave re-heating has been an important factor with respect to added convenience. The major trends in the development of fruit and vegetable heat processed products in recent years are shown in table; the number of new fruit and vegetable products.

***Numbers of New Fruit and Vegetable Products***

| | 1990 | 1991 | 1992 Jan-June |
|---|---|---|---|
| Frozen vegetable products | 66 | 95 | 21 |
| Chilled vegetable products | 76 | 81 | 78 |
| Heat processed vegetables | 51 | 60 | 38 |
| Heat processed fruits | 13 | 14 | 5 |
| Fruit juices and drinks | 73 | 83 | 46 |
| Potato crisps | 32 | 33 | 16 |

***Source:*** C. Dennis (1993)

New product development in the fruit and vegetable sector is most important in meeting the continued challenge of providing the consumer with choice and high quality products.

## A FRESH LOOK AT DRIED FRUIT

New fruit varieties and advance in drying technologies are putting a fresh twist on dried fruit applications. Fruits that have been introduced to the drying process include cranberries, blueberries, cherries, apples, raspberries and strawberries - not to mention the traditional mainstays of raisins, dates, apricots, peaches, prunes and figs.

Perceived as a "value-added" ingredient, dried fruit adds flavour, colour, texture and diversity with little alteration to an existing formula. The growing interest in ethnic cuisines in U.S.A. and the change to a more healthy way of eating, has also moved dried fruit considerably closer to the mainstream.

Found primarily in the baking industry, dried fruit is coming into its own in various food products, including entrees, side dishes and condiments. Compotes, chutneys, rice and grain dishes, stuffings, sauces, breads, muffins, cookies, deserts, cereals and snacks are all food categories encompassing dried fruit.

Since some dried fruit is sugar infused (osmotic drying), food processors can decrease the amount of sugar in formula-this is especially the case in baked products. Processors are making adjustments in moisture content of the dried fruit so that a varied range is available for different applications. An added bonus is dried fruits' shelf stability (a shelf life of at least 12 months). Dried fruit is more widely available in different forms, including whole dried, cut, diced and powders.

Citric acid and its use in fruit and vegetable processing.

Citric acid may be considered as "Nature's acidulant".

It is found in the tissues of almost all plants and animals, as well as many yeasts and moulds.

Commercially citric acid is manufactured under controlled fermentation conditions that produce citric acid as a metabolic intermediate from naturally-occurring yeasts, moulds and nutrients. The recovery process of citric acid is through crystallization from aqueous solutions.

Citric acid is widely used in carbonated and still beverages, to impart a fresh-fruit "tanginess". Citric acid provides uniform acidity, and its light fruity character blends well and enhances fruit juices, resulting in improved palatability. The amount of citric acid used depends on the particular desired flavour (*e.g.*, High-acid: lemonade; Medium-acid: orange, punch, cherry; Low-acid: strawberry, black cherry, grape).

Sodium citrate is often added to beverages to mellow the tart taste of high acid concentrations. It provides a cool, distinctive smooth taste and masks any bitter aftertaste of artificial sweeteners. In addition, it serves as a buffer to stabilise the pH at the desired level. The high water solubility of citric acid (181 g/100 ml) makes it an ideal additive for fountain fruit syrups and beverages concentrates as a flavour enhancer and microbial growth inhibitor (preferably at $pH < 4.6$).

In processed fruits and vegetables, citric acid performs the following functions:

a. It reduces heat-processing requirements by lowering pH: inhibition of microbial growth is a function of pH and heat treatment. Higher heat exposure and lower pH result in greater inhibition. Thus the use of citric acid to bring pH below 4.6 can reduce the heating requirements. In canned vegetables, citric acid usage is greatest in tomatoes, onions and pimentos. For tomato packs, the National Canners Association recommends a pH of 4.1 to 4.3. In general, 0. 1% citric acid will reduce the pH of canned tomatoes by 0.2 pH units.

b. Optimise flavour: citric acid is added to canned fruits to provide for adequate tartness. Recommended usage level is generally less then 0.15%.

c. Supplement antioxidant potential: citric acid is used in conjunction with antioxidants such as ascorbic and erythorbic acids, to inhibit colour and flavour deterioration caused by metal-catalysed enzymatic oxidation. Recommended usage levels are generally 0.1% to 0.3% with the antioxidant at 100 to 200 ppm.

d. Inactivate undesirable enzymes: oxidative browning in most fruits and vegetables is catalysed by the naturally present polyphenol oxidase. The enzymatic activity is strongly dependent on pH.

Addition of citric acid to reduce pH below 3 will result in inactivation of this enzyme and prevention of browning reactions.

**Cherry and Apricot Oils are Safe for Food Use**

The oils obtained by cold pressing the kernels of the cherry (Prunus cerasus) and the apricot (Prunus armeniaca) have been declared acceptable for food use by the UK Ministry of Agriculture, Fisheries & Food's Advisory Committee on Novel Foods and Processes (ACNFP) by June 1993.

In its assessments of the safety in use of the cherry and apricot kernel oils, the ACNFP consider specifications that included data on fatty acid composition, the presence of natural antioxidants and the content of cyanide, mycotoxins and heavy metals.

The Committee says that it gave particular consideration to the possible presence in the oils of the cyanogenic glucoside amygdalin, from which cyanide is released by enzymic action when the kernels of cherry and apricot are crushed. Amygdalin was found to be absent from the cherry and apricot kernel oils.

The ACNFP is satisfied that there are no food safety reasons why the use of cherry and apricot kernel should not be acceptable provided there is compliance with the specifications.

***Specification of Purity for Cherry and Apricot Kernel Oils as Determined by UK ACNFP***

| | Cherry | Apricot |
|---|---|---|
| Contaminants limits: | | |
| Heavy metals (total) | 0.5 mg/kg | 0.5 mg/kg |
| Aflatoxins (total) | 4.0 g/kg | 4.0 g/kg |
| Cyanide | 0.15 mg/kg | 0.15 mg/kg |
| Pesticide residues | 0.01 mg/kg | 0.01 mg/kg |
| Tocopherols: | | |
| alpha/delta/gamma (mg/kg) | 356-886 | 569-899 |

***Source:*** Anon. (1993)

The oils are obtained by the mechanical mincing and cold pressing of kernels extracted from cleaned cherry or apricot stones. After filtering, the oils are stored and are to be sold in a raw, unrefined state. The cherry and apricot kernel oils are high unsaturated and are expected to be used as speciality oils for salad dressings, baking and shallow frying applications.

## THE USE OF FRUIT JUICES IN CONFECTIONERY PRODUCTS

During the last decade, the concept of fruit juices has gained immensely on consumer popularity. The majority of new non-alcoholic and alcoholic fruit drink products were a combination of syrups, fruit juices and flavours. The confectionery industry followed suit and new products incorporated fruit juices as part of their confectionery formulations and processes. Fruit juice concentrates of high solids are often used instead of normal or single-fold juices.

Juice concentrates are made of pure fruit juices. The process starts with pressing fruits and obtaining pure fruit juice; this is stabilised by heat treatment which inactivates enzymes and microorganisms. The next processing step is concentration under vacuum up to 40-65° Brix or 4-7 fold. The concentrates are then blended for standardisation and stored.

These fruit juice concentrates are often further stabilised by the addition of sodium benzoate and potassium sorbate and are usually stored away from light and are refrigerated or frozen.

Depectinised fruit juices are also used to prevent foaming in confectionery processes and are essential for use in clear beverage products. Fruit juice concentrates which are depectinised, and have added preservatives are called stabilised, clarified, fruit juice concentrates.

Fruit juices are used in confectionery products in conjunction with natural and artificial flavours which provides intense flavour impact and are cost-effective for a confectionery product.

The traditional concern in using fruit juice concentrates in confectionery applications has been the effect of the natural acids on the finished product, particularly the formation of invert sugar during processing.

This is a logical concern since concentrates contain differing amounts and types of acids. For example: apple, cherry, strawberry and other berries contain primarily malic acid. Grapes mainly contain tartaric acid. Cranberry is high in quinic acid. Citrus fruits and pineapple contain differing amounts of citric acid. The concentrates, when used, are normally buffered to a pH of 5-7 with sodium hydroxide.

In formulating products with fruit juice concentrates, the solids of the concentrate are considered as mostly reducing sugars and a reduction in corn syrup is made to compensate for equivalent amount of reducing sugar being added in the concentrate. The exact replacement can be determined by measuring the D.E. of the concentrate to be added. In formulations when small amounts of concentrate are used (less than 1%), no adjustment is made since the reducing sugar contribution of the concentrate is not significant.

Fruit juice concentrates can also be used to provide a source of natural colour, in particular red colour. Grape, raspberry, cherry, strawberry and cranberry concentrates in small amounts are very effective in colouring cream centres.

The inclusion of fruit juices in confectionery products is now left up to the imagination of the manufacturer. These products must, of course, hold up to the standards of flavour integrity, and product excellence, during the shelf-life of these products.

## VEGETABLE SPECIFIC PROCESSING TECHNOLOGIES

### Vegetables Varieties

Vegetable processors must appreciate the substantial differences that varieties of a given vegetable will possess. In addition to variety and genetic strain differences with respect to weather, insect and disease resistance, varieties of a given vegetable will differ in size, shape, time of maturity, and resistance to physical damage.

Varietal differences then further extend into warehouse storage stability, and suitability for such processing methods as canning, freezing, pickling or drying. A variety of peas that is suitable for canning may be quite unsatisfactory for freezing and varieties of potatoes that are preferred for freezing may be less satisfactory for drying or potato chip manufacture.

This should be expected since different varieties of a given vegetable will vary somewhat in chemical composition, cellular structure and biological activity of their enzyme system.

### Harvesting and Pre-processing

When vegetables are maturing in the field they are changing from day to day. There is a time when the vegetable will be at peak quality from the stand-point of colour, texture and flavour. This peak quality is quick in passing and may last only a day. Harvesting and processing of several vegetables, including tomatoes, corn and peas are rigidly scheduled to capture this peak quality. After the vegetable is harvested it may quickly pass beyond the peak quality condition. This is independent of microbiological spoilage; these main deteriorations are related to:

(a) loss of sugars due to their consumption during respiration or their conversion to starch; losses are slower under refrigeration but there is still a great change in vegetable sweetness and freshness of flavour within 2 or 3 days;

(b) production of heat when large stockpiles of vegetables are transported or held prior to processing.

At room temperature some vegetables will liberate heat at a rate of 127,000 kJ/ton/day; this is enough for each ton of vegetables to melt 363 kg of ice per day. Since the heat further deteriorates the vegetables and speeds microorganisms growth, the harvested vegetables must be cooled if not processed immediately.

But cooling only slows down the rate of deterioration, it does not prevent it, and vegetables differ in their resistance to cold storage. Each vegetable has its optimum cold storage temperature which may be between about 0-100 C (32-500 F).

(c) the continual loss of water by harvested vegetables due to transpiration, respiration and physical drying of cut surfaces results in wilting of leafy vegetables, loss of plumpness of fleshy vegetables and loss of weight of both.

Moisture loss cannot be completely and effectively prevented by hermetic packaging. This was tried with plastic bags for fresh vegetables in supermarkets but the bags became moisture fogged, and deterioration of certain vegetables was accelerated because of buildup of CO2 and decrease of oxygen in the package.

It therefore is common to perforate such bags to prevent these defects as well as to minimise high humidity in the package which would encourage microbial growth.

Shippers of fresh vegetables and vegetable processors, whether they can, freeze, dehydrate, or manufacture soups or ketchup, appreciate the instability and perishability of vegetables and so do everything they can to minimise delays in processing of the fresh product. In many processing plants it is common practice to process vegetables immediately as they are received from the field.

To ensure a steady supply of top quality produce during the harvesting period the large food processors will employ trained field men; they will advise on growing practices and on spacing of plantings so that vegetables will mature and can be harvested in rhythm with the processing plant capabilities. This minimises stockpiling and need for storage.

Cooling of vegetables in the field is common practice in some areas. Liquid nitrogen-cooled trucks may next provide transportation of fresh produce to the processing plant or directly to market. Upon arrival of vegetables at the processing centre the usual operations of cleaning, grading, peeling, cutting and the like are performed using a moderate amount of equipment but a good deal of hand labour also still remains.

Reception: This covers qualitative and quantitative control of delivered vegetables. The organoleptic control and the evaluation of the sanitary state, even if they are very important steps in vegetables' characteristics assessment, cannot establish their technological value.

On the other hand, laboratory controls do not precisely establish their technological properties because of the difficulty in putting into showing some deterioration when using rapid control methods.

One correct method of vegetable quality appraisal is their overall evaluation based on the whole complex of data that can be obtained by combining an extensive organoleptic evaluation with simple analysis that can be performed rapidly in plant laboratory. These analysis can be:

a. refractometric extract (tomatoes, fruit, etc.);

b. specific weight (potatoes, peas, etc.);

c. consistency (measured with tenderometers, penetrometers, etc.);

d. boiling tests, etc.

Temporary storage.

This step should be as short as possible and better completely eliminated. Vegetables can be stored in:

a. simple stores, without artificial cooling;

b. in refrigerated stores; or, in some cases,

c. in silos (potatoes, etc.).

Simple stores should be covered, fairly cool, dry and well ventilated but without forced air circulation which can induce significant losses in weight through intensive water evaporation; air relative humidity should be at about 70-80%.

Refrigerated storage is always preferable and in all cases a processing centre needs a cold room for this purpose, adapted in volume I capacity to the types and quantities of vegetables (and fruits) that are further processed.

***Washing:*** Washing is used not only to remove field soil and surface microorganisms but also to remove fungicides, insecticides and other pesticides, since there are laws specifying maximum levels of these materials that may be retained on the vegetable; and in most cases the allowable residual level is virtually zero. Washing water contains detergents or other sanitisers that can essentially completely remove these residues.

The washing equipment, like all equipment subsequently used, will depend upon the size, shape and fragility of the particular kind of vegetable:

- flotation cleaner for peas and other small vegetables;
- rotary washer in which vegetables are tumbled while they are sprayed with jets of water; this type of washer should not be used to clean fragile vegetables;

***Sorting:*** This step covers two separate operations:

(a) Removal of non-standard vegetables (and fruit) and possible foreign bodies remaining after washing;

(b) Quality grading based on variety, dimensional, organoleptical and maturity stage criterion.

***Skin Removal/peeling:*** Some vegetables require skin removal. This can be done in various ways.

(a) Mechanical

This type of operation is performed with various types of equipment which depend upon the result expected and the characteristics of the fruit and vegetables, for example:

i. a machine with abrasion device (potatoes, root vegetables);

ii. equipment with knives (apples, pears, potatoes, etc.);

iii. equipment with rotating sieve drums (root vegetables). Sometimes this operation is simultaneous with washing (potatoes, carrots, etc.) or preceded by blanching (carrots).

(b) Chemical

Skins can be softened from the underlying tissues by submerging vegetables in hot alkali solution. Lye may be used at a concentration of about 0.5-3%, at about 93° C (2000 F) for a short time period (0.5-3 min). The vegetables with loosened skins are then conveyed under high velocity jets of water which wash away the skins and residual lye.

In order to avoid enzymatic browning, this chemical peeling is followed by a short boiling in water or an immersion in diluted citric acid solutions.

It is more difficult to peel potatoes with this method because it is necessary to dissolve the cutin and this requires more concentrated lye solutions, up to 10%.

(c) Thermal

***Wet Heat (steam):*** Other vegetables with thick skins such as beets, potatoes, carrots and sweet potatoes may be peeled with steam under pressure (about 10 at) as they pass through cylindrical vessels. This softens the skin and the underlying tissue. When the pressure is suddenly released, steam under the skin expands and causes the skin to puff and crack. The skins are then washed away with jets of water at high pressure (up to 12 at).

***Dry Heat (flame):*** Other vegetables such as onions and peppers are best skinned by exposing them to direct flame (about 1 min at 1000° C) or to hot gases in rotary tube flame peelers. Here too, heat causes steam to develop under skins and puff them so that they can be washed away with water.

Manual peeling only use when the other methods are impossible or sometimes as a completion of the other three ways. Average losses at this step are given in Table.

***Losses at Vegetable Peeling, in %***

| | Peeling methods | | |
|---|---|---|---|
| Vegetables | Manual | Mechanical | Chemical |
| Potatoes | 15-19 | 18-28 | - |
| Carrots | 13-15 | 16-18 | 8-10 |
| Beets | 1416 | 13-15 | 9-10 |

***Size Reduction:*** This step is applied according to specific vegetable and processing technology requirements.

***Blanching:*** The special heat treatment to inactivate enzymes is known as blanching. Blanching is not indiscriminate heating. Too little is ineffective, and too much damages the vegetables by excessive cooking, especially where the fresh character of the vegetable is subsequently to be preserved by processing.

This heat treatment is applied according to and depends upon the specificity of vegetables, the objectives that are followed and the subsequent processing / preservation methods.

Two of the more heat resistant enzymes important in vegetables are catalase and peroxidase. If these are destroyed then the other significant enzymes in vegetables also will have been inactivated. The heat treatment to destroy catalase and peroxidase in different vegetables are known, and sensitive chemical tests have been developed to detect the amounts of these enzymes that might survive a blanching treatment. Catalase and peroxidase inactivation tests are presented in section 9.2.9.

Because various types of vegetables differ in size, shape, heat conductivity, and the natural levels of their enzymes, blanching treatments have to be established on an experimental basis. As with sterilisation of foods in cans, the larger the food item the longer it takes for heat to reach the centre. Small vegetables may be adequately blanched in boiling water in a minute or two, large vegetables may require several minutes.

Blanching as a unit operation is a short time heating in water at temperatures of 100° C or below. Water blanching may be performed in double bottom kettles, in special baths with conveyor belts or in modern continuous blanching equipment.

In order to reduce losses of hydrosoluble substances (mineral salts, vitamins, sugars, etc.) occurring during water blanching, several methods have been developed:

- temperature setting at 85-95° C instead of 100° C;
- blanching time has to be just sufficient to inactivate enzymes catalase and peroxidase;
- assure elimination of air from tissues.

An illustration of blanching parameters is seen in Table.

***Blanching Parameters for Some Vegetables***

| *Vegetables* | *Temperature, °C* | *Time, min.* |
|---|---|---|
| Peas | 85-90 | 2-7 |
| Green beans | 90-95 | 2-5 |
| Cauliflower | Boiling | 2 |
| Carrots | 90 | 3-5 |
| Peppers | 90 | 3 |

Steam heat treatment can also be applied instead of water blanching as a preliminary step before freezing or drying, as long as the preservation method is only used for enzyme inactivation and not to modify consistency.

For drying, the vegetables are conveyed directly from steaming equipment to drying installations without cooling.

Vegetable steaming is carried out in continuous installations with conveyer belts made from metallic sieves. Cooling of vegetables after water blanching or steaming is performed in order to avoid excessive softening of the tissues and has to follow immediately after these operations; one exception is the case of vegetables for drying which can be transferred directly to drying equipment without cooling.

Natural cooling is not recommended because is too long and generates significant losses in vitamin C content. Cooling in pre-cooled air (from special installations) is sometimes used for vegetables that will be frozen

Cooling in water can be achieved by sprays or by immersion; in any case the vegetables have to reach a temperature value under 37° C as soon as possible. Too long a cooling time generates supplementary losses in valuable hydrosoluble substances; in order to avoid this, the temperature of the cooling water has to be as low as possible.

***Canning:*** Large quantities of vegetable products are canned. A typical flow sheet for a vegetable canning operation (which also applies to fruit for the most part) covers some food process unit operations performed in sequence: harvesting; receiving; washing; grading; heat blanching; peeling and coring; can filling; removal of air under vacuum; sealing/closing, retorting/heat treatment; cooling; labelling and packing. The vegetable may be canned whole, diced, puréed, as juice and so on.

## On-line Simplified Methods for Enzyme Activity Check

*Peroxidase Test:*

(a) *Solutions:* In order to check the peroxidase activity two solutions have to be prepared:

- 1% guaiacol in alcohol solution (1 g guaiacol is dissolved in about 50 cm$^3$ of 96% ethylic alcohol and then this preparation is brought to 100 cm° with the same solvent);
- peroxide solution 0.3% (1 cm$^3$ perhydrol is brought to 100 cm$^3$ with distilled water.

(b) *Sampling:* From various parts of the material samples are taken (about 20-30 pieces, etc.); the material is then crushed in a laboratory bowl in order to obtain an average sample.

(c) *Check:* Prom the average sample, 10-20 g of material is taken in a medium capacity test tube; on this sample are poured: 20 $cm^3$ distilled water; 1 $cm^3$ of 1% guaiacol solution; 1.6 $cm^3$ of peroxide solution.

The contents of the test tube is shaken well. The gradual appearance of a weak pink colour indicates an incomplete peroxidase inactivation - reaction slightly positive. If there are no tissue colour modifications after 5 minutes, the reaction is negative and the enzymes have been inactivated. As an orientative check it is also possible to simply pour a few drops of 1% guaiacol solution and 0.3% peroxide solution directly on blanched and crushed vegetables. A rapid and intensive brown-reddish tissue colouring indicates a high peroxidase activity (positive reaction).

***Catalase Test:*** In order to identify the catalase enzyme activity, 2 g of dehydrated vegetables are well crushed and mixed with about 20 $cm^3$ of distilled water. After 15 min softening, 0.5 $cm^3$ of a 0.5% or 1% peroxide solution is poured on prepared vegetables. In the presence of catalase, a strong oxygen generation is observed for about 2-3 minutes. These tests are of a paramount importance in order to determine the vegetable blanching treatments (temperature and time); incomplete enzyme inactivation has a negative effect on finished product quality. For cabbage catalase inactivation by blanching is sufficient; blanching further to peroxidase inactivation would have negative effects on product quality and even complete browning. For all other vegetables and for potatoes, both tests MUST be negative, for catalase and for peroxidase.

## FRESH VEGETABLE STORAGE

The vegetables can be stored, in some specific natural conditions, in fresh state, that is without significant modifications of their initial organoleptic properties. Fresh vegetable storage can be short term; this was briefly covered under temporary storage before processing. Also fresh vegetable storage can be long term during the cold season in some countries and in this case it is an important method for vegetable preservation in the natural state. In order to assure preservation in long term storage, it is necessary to reduce respiration and transpiration intensity to a minimum possible; this can be achieved by:

a. maintenance of as low a temperature as possible (down to 0° C),

b. air relative humidity increased up to 85-95 % and

c. $CO_2$ percentage in air related to the vegetable species.

Vegetables for storage must conform to following conditions: they must be of one of the autumn or winter type variety; be at edible maturity without going past this stage; be harvested during dry days; be protected from rain, sun heat or wind; be in a sound state and clean from soil; be undamaged. From the time of harvest and during all the period of their storage vegetables are subject to respiration and transpiration and this is on account of their reserve substances and water content. The more the intensity of these two natural processes are reduced, the longer sound storage time will be and the more losses will be reduced. For this reason, vegetables have to be handled and transported as soon as possible in the storage conditions (optimal temperature and air relative humidity for the given species). Even in these optimal conditions storage will generate losses in weight which are variable and depend upon the species. Some optimal storage conditions are shown in table.

***Optimal Conditions for Fresh Vegetable Storage***

| Vegetables | Storage conditions | |
|---|---|---|
| | Temperature, °C | Relative humidity, % |
| Potatoes | +1...+3 | 85-90 |
| Carrots | 0 ... +1 | 90-95 |
| Onions | 0 ... +1 | 75-85 |
| Leeks | 0 ... +0.5 | 85-90 |
| Cabbage | -1 ... 0 | 90-97 |
| Garlic | 0 ... +1 | 85-90 |
| Beets | 0 ... +1 | 90-95 |

# 6

# PRESERVATION TECHNIQUES FOR VEGETABLES

## Technology for Vegetable Powder Processing

This technology has been developed in recent years with applications mainly for potatoes (flour, flakes, granulated), carrots (powder) and red tomatoes (powder). In order to obtain these finished products there are two processes:

(a) Drying of vegetables down to a final water content below 4% followed by grinding, sieving and packing of products;

(b) Vegetables are transformed by boiling and sieving into purees which are then dried on heated surfaces (under vacuum preferably) or by spraying in hot air.

Industrial installations that can be used for these products and technological data are summarised below:

- Dryers with plates under vacuum are equipped with plates heated with hot water. Stainless steel plates containing the puree to be dried are placed on them. Process conditions are at low residual pressure (about 10-20 mm Hg) and a product temperature of 50-70° C. This equipment is discontinuous but easy to operate.
- Drum dryers have one or two drums heated with hot water or steam as heating elements. Feeding is continuous between the two drums which are rotating in reverse direction (about 2-6 rotations per minute) and the distance of which is adjustable and determines

the thickness of layer to be dried, he product is dried and removed by mechanical means during rotation.

- Drying installations by spraying in hot air; the product is introduced in equipment and sprayed by a special device in hot air. Drying is instantaneous (1/50 s) and therefore can be carried out at 130-15O° C.

**Packing and Storage of Dried and Powdered Vegetables**

Dried vegetables can suffer significant modifications that bring about their deterioration during storage. The factors that determine these degradations impose at same time the type of packaging materials and storage conditions for packaged products.

The main factor in maintaining the quality of dried products is to follow the maximum moisture contents that have to be as close as possible to the limits indicated.

The moisture content of dried vegetables is not constant because of their hygroscopicity and is always in equilibrium with relative humidity of air in storage rooms. Technical solutions for maintaining a low dehydrated products moisture are:

(a) storage in stores with air relative humidity below 78%;

(b) use packages that are water vapour proof. The most efficient packages are tin boxes or drums (mainly for long term storage periods); combined packages (boxes, bags, etc.) from complexes (carton with metallic sheets, plastic materials, etc.) mainly for small packages. One solution for some dried vegetables may be the use of waterproof plywood drums.

Modern solutions are oriented not only to the maintaining product moisture during storage but also reducing this parameter by the use of desiccants (substances which absorb moisture) introduced in packages, hermetically closed.

A desiccant in current use is calcium oxide. Granulated calcium oxide is introduced in small bags from a material which is permeable to water vapour but which does not permit the desiccant to escape into products.

With desiccants, product moisture can be reduced to even below 4%, and this inhibits or reduces the biochemical and microbiological processes during storage.

Another factor that can deteriorate dried/dehydrated vegetables is atmospheric oxygen through the oxidative phenomena that it produces. In order to eliminate the action of this agent some packing methods under vacuum or in inert gases (carbon dioxide or nitrogen) are in use, applied mainly for packing dried carrots in order to avoid beta-carotene oxidation in beta-ionone (foreign smell, discoloration, etc.). In order to avoid the action of oxygen it is also possible to add ascorbic acid as antioxidant (for example in carrot powder).

Sun or artificial light action on dehydrated vegetables generally causes discoloration which can be avoided by using opaque packaging materials.

Dehydrated vegetable compression (especially for roots) to form blocks with a weight of 50-600 g, is practiced sometimes; it has as advantages the reduction of evaporation surface and contact with atmospheric oxygen and volume reduction. Dehydrated vegetables are compressed at about 300 at. Compressed blocks are packaged in heat sealed plastic materials.

Storage temperature has an important role because this reduces or inhibits the speed of all physico-chemical, biochemical and microbiological processes, and thus prolongs storage period. The storage temperature should be below 25° C (and preferably 15° C); lower temperatures (0-10° C) help maintain taste, colour and water rehydration ratio and also, to some extent, vitamin C.

## POTATO CRISP/CHIP PROCESSING

The most important steps involved in potato crisps processing are:

1. Selecting, procuring and receiving potatoes
2. Storage of potato stock under optimum conditions
3. Peeling and trimming the tubers
4. Slicing

5. Frying in oil
6. Salting or applying flavoured powders
7. Packaging

***Moisture and Shipping Factors for Some Dehydrated Vegetables***

| Product | Form/cut | Moisture % | Weight kg/m³ |
|---|---|---|---|
| Bean (green) | 20 nun cut | 5 | 1.6 |
| Bean (lima) | 5 | 3.3 | |
| Beet | 6 mm strips | 5 | 1.6-1.9 |
| Cabbage | 6-12 mm shreds | 4 | 0.7-0.9 |
| Carrots | 5-8 mm strips | 5 | 3-5 |
| Celery | Cut | 4 | |
| Garlic | Cloves | 4 | |
| Okra | 6 mm slices | 8 | |
| Onion | Slices | 4 | 0.4- 0.6 |
| Pea (fresh) | Whole | 5 | 3.4 |
| Pepper (hot) | Ground | 5 | |
| Pepper (sweet) | 5 mm strips | 7 | |
| Potato (Irish) | 5-8 mm strips | 6 | 2.9-3.2 |
| | Diced | 5 | 3.3-3.6 |
| Tomato | 7-10 mm slices | 35 | |

## SELECTION AND STORAGE

It is important to select potatoes of high specific gravity since this characteristic indicates superior yield and lower oil absorption. It is even more important to select potatoes with low reducing sugar contents or to store them at temperatures conducive to the minimising of these substances.

Sprouting and fungal damage must also be minimised by the storage conditions.

### Peeling

The ideal peeling operation should only remove a very thin outer layer of the potato, leaving no eyes, blemishes, or other material for later removal by hand trimming. It should not significantly change the physical or chemical characteristics of the remaining tissue.

Preferably peeling should use small amount of water and result in minimal effluent; compromises will have to be made in all of these aspects of peeling. First, the potatoes are thoroughly washed, not only for sanitary reasons, but also to prevent dirt of grit from abrading the equipment the tubers will dater contact. Washing may take place in streams, as the potatoes are being conveyed by water streams, or in equipment provided with means for scrubbing the potato with brushes or rubber rolls.

In barrel-type washers, potatoes are cleaned by being tumbled and rubbed against each other and against the sides of the barrel while they are immersed in, or sprayed with, water. After washing, the potatoes are allowed to drain, usually on mesh conveyors, and they travel over an inspection belt where foreign material and defective tubers are removed. The more common peeling methods are abrasion, lye immersion, and steam.

Abrasion peelers which may be either batch or continuous, use disks or rollers coated with grit to grind away the potato surface. An important design feature is to ensure that all surfaces of the tuber are equally exposed to the rasping action. The peel fragments are flushed out of the unit by water sprays.

Such systems work best with uniform, round, undamaged potatoes. Some of the advantages of abrasion peelers are their simplicity, compactness, low cost, and convenience. They are particularly suitable for peeling potatoes intended for chipping, since they do not chemically alter the surface layers. About 10% of the original tuber weight is lost through abrasion peeling prior to chipping.

## Slicing

The peeled potatoes are cut into slices from 1/15 to 1/25 in. by rotary slicers. Centrifugal force presses the tuber against stationary gauging shoes and knives. Thickness is varied, not only to meet consumer preferences, but also to fit the condition of the tubers and the frying temperature and time. Slices produced at any one time must be very uniform in thickness, however, in order to obtain uniformly coloured chips. Slices with rough or torn surfaces lose excess solubles from ruptured cells and absorb larger amounts of fat.

It is necessary to remove the starch and other material released from the cut cells from the surface of slices so that the slices will separate readily and completely during frying. The slices are washed in stainless steel wire-mesh cylinders or drums rotating in a rectangular stainless steel tank. After washing and an additional rinse in similar equipment, the potatoes may or may not be dried.

## Frying

The capacity of the fryer is generally the limiting factor in the process line. Most manufacturers currently use continuous fryers but some batch equipment is still employed.

Modern continuous fryers have the following essential elements: (1) a tank of hot oil in which the chips are cooked; (2) a means for heating and circulating the oil; (3) a filter for removing particles from oil; (4) a conveyor to carry chips out of the tank; (5) a reservoir in which oil is heated for adding to the circulating frying oil and (6) vapour-collecting hoods above the tank. Temperatures normally used are from 350 to 375° F at the receiving end and 320 to 345° F at the exit end.

The oil used for deep-fat frying of potato chips has two functions:

(i) it serves as a medium for transferring heat from a thermal source to the tuber slices;

(ii) it becomes an ingredient of the finished product.

Use of highly refined oil is of great importance in flavour and stability of the crisps. Flavour, texture, and appearance are affected both by the amount of oil absorbed and its characteristics as it exists in the crisp (*i.e.* not necessarily its initial chemical and physical parameters).

Oils change continuously during the frying process but the heat abuse resulting from the crisp cooking is relatively mild.

Temperatures rarely rise above 385° F at any point. Better control over crisp colour could be obtained if the final stage of moisture removal could be achieved without the browning reaction that always accompanies it in the frying process.

Crisps may be sorted for size after frying, with the larger crisps being diverted to the bulk packs and larger pouches and the smaller pieces used for vending machine packs and other individual service containers. Potato crisp sizing is also accomplished by separating the peeled potatoes into large and small sizes, which are then sliced and fried separately.

The crisps are salted immediately after they leave the fryer. It is important that the fat be liquid at this point to cause maximum adherence of the granules. Powders containing barbecue spices, cheese, or other speciality materials may be added at this point. The salt may contain added enrichment materials or antioxidants.

After salting, the crisps pass on to a conveyor belt where they are visually inspected and off-colour material is removed. I the crisps are allowed to cool before packaging, better adherence of salt and flavour powders is obtained.

Some consumers prefer the hard, curled-up crisp that is characteristic of the hand-kettle type of operation. The special flavour of the hand-kettle crisp is said to be due, at least partly, to the starch retained on the cut surfaces of the potato slices as a result of the omission of a washing process after slicing. Starch-covered slices tend to stick together in the fryer so it is necessary to use devices to prevent clumping.

The principal factors affecting potato crisp acceptability are piece size, colour, and of course, flavour. These factors are controllable primarily by selection of the raw material, adjustment of processing conditions, and packaging.

**Storage Stability**

If the frying oil is stabilised and has not deteriorated through use, and if the packaging is opaque and has a low moisture vapour transmittance rate (MVTR), a shelf-life of 4-6 weeks should be achieved when crisps are stored at temperatures of about 70° F.

Once potato crisps are in the bag, the three forms of quality loss which have the greatest effect on consumer acceptance are breakage, absorption of moisture with loss of crispiness, and fat oxidation leading to development of rancid odours.

The mechanical abuse causing breaking of the crisps can be partially prevented by using stiff packaging material, making the package "plump" with contained air, and avoiding crushing in the shipping case.

Absorption of moisture is prevented largely by proper choice of packaging material. Cellophane coated with various moisture barriers has proved to be a satisfactory pouch films for the relatively short shelf-life expected (generally stated to be 4-6 weeks).

Light (especially fluorescent light) accelerates oxidation, so that opaque packaging material must be used to obtain maximum shelf-life.

Potato crisps are considered commercially unacceptable when they have a moisture content above 3%, which is in equilibrium with a relative humidity of about 32%. The containers should have a high degree of resistance to moisture-vapour transfer.

If pouches are used, foil-containing films are preferable, since they not only resist moisture-vapour transfer but reflect light.

## VEGETABLE JUICES AND CONCENTRATED PRODUCTS

### Vegetable Juices

Vegetable juices are natural products constituted from cellular juice and a part of crushed pulp, from the tissues of some vegetables. These juices contain all valuable substances from the vegetables: vitamins, sugars, acids, mineral salts and pectic substances. The most important of these products is tomato juice; in a lower proportion there are also other juices (carrots, beet, sauerkraut, etc.).

***Tomato Juice:*** This product is characterised not only by its organoleptical properties (taste, colour, flavour) but also by its vitamin content close to those of fresh tomatoes. Modem technology is oriented to a maximum maintenance of organoleptic properties and of vitamin content.

At same time, it is important to assure juice uniformity by avoiding cellulosic particle sedimentation. Juice stability is assured by a flash pasteurization which assures the destruction of natural micro-flora, while keeping the initial properties.

The modern technological flow-sheet covers the following main operations: Pre-washing is carried out by immersion in water, cold or heated up to 50° C (possibly with detergents to eliminate traces of pesticides). This operation is facilitated by bubbling compressed air in the immersion vessel/equipment.

Washing is performed with water sprays, which in modern installations have a pressure of 15 at or more.

Sorting/Control on rolling sorting tables enables the removal of non-standard tomatoes - with green parts, yellow coloured, etc.

## Crushing in Special Equipment

Preheating at 55-60° C facilitates the extraction, dissolves pectic substances and contributes to the maintaining of vitamins and natural pigments. In some modern installations, this step is carried out under vacuum at 630-680 mm Hg and in very short time.

Extraction of juice and part of pulp (maximum 80%) is performed in special equipment / tomato extractors with the care to avoid excessive air incorporation. In some installations, as an additional special care, a part of pulp is removed with continuous centrifugal separators.

De-aeration under high vacuum of the juice brings about its boiling at 35-40° C.

Homogenisation is done for mincing of pulp particles and is mandatory in order to avoid future potential product "separation" in two layers.

Flash Pasteurization at 130-150° C, time = 8-12 see, is followed by cooling at 90° C, which is also the filling temperature in receptacles (cans or bottles).

## ASEPTIC FILLING

Closing Of Receptacles is followed by their inversion for about 5 to 7 minutes. Cooling has to be carried out intensely. Full cans do not need further pasteurization because the bacteria that have potentially contaminated the tomato juice during filling are easily destroyed at 90° C due to natural juice acidity. For bottles, it may be possible to avoid further sterilisation if the following conditions can be respected: washing and sterilising of receptacles, cap sterilisation (with formic acid), filling and capping under aseptic conditions, in a space with UV lamps. In so far as this is quite difficult to achieve it may be necessary to submit bottles to a pasteurization in water baths. The main characteristics of high quality tomato juice are:

- natural red colour;
- taste and flavour of fresh tomatoes;
- uniformity (without pulp sedimentation);
- total soluble solids: 6% minimum;
- total soluble substances (by refractometer): 5% minimum;
- vitamin C: 15 mg/100ml minimum.

In traditional processes it is recommended to:

- thoroughly wash and rinse the empty receptacles (including jar caps / covers and bottle crown corks) and then "sterilize" by keeping in boiling water for 30 mini
- add salt and lemon juice to the prepared receptacles just before filling;
- pasteurize closed glass receptacles (bottles or jars) according to conditions recommended in technological flow-sheets and which is summarised as follows:

| Receptacle size | Pre-heating | Time of pasteurization |
|---|---|---|
| 0.33 1 | 60° C | 40 minutes |
| 0.501 | 60° C | 45 minutes |
| 0.66 1 | 60° C | 55 minutes |
| 0.751 | 60° C | 60 minutes |
| 1.0 litre | 60° C | 70 minutes |

***Carrot Juice:*** This product represents an important dietetic product due to its high soluble pectin content. Technological flow-sheet is oriented to the maintaining of as high as possible a pectin content and covers the following steps:

*Pre-washing*

*Cleaning*

*Washing*

*Blanching in steam for 20 minutes*

*Grating*

*Pressing*

Juice in the pressed juice will then be incorporated 25% of grated carrot (non pressed)

Homogenisation in colloidal mills

Acidification with 0.25% citric or tartric acid

De-aeration

Filling in receptacles (bottles or tinplate cans)

Airtight sealing

Pasteurization at 100° C for 30 minutes.

The main characteristics of a good quality carrot juice:

- uniformity (no separation in layers occurs during storage);
- good orange colour;
- pleasant taste, close to fresh carrot taste;
- total soluble solids: 12 %;
- total sugar content: 8%;
- beta-carotene: 1.3 mg/100 ml;
- soluble pectin: 0.4 %.

***Red Beet Juice:*** The product is obtained following this technological flow-sheet: washing, cleaning, steam treatment/ steaming (30-35 min at 1050 C), pressing, strain through small hole sieve, filling in receptacles, tight sealing/closing, sterilisation (25 min at 1 15° C). In order to improve taste, the juice is acidified with 0.3% citric or tartric acid.

***Sauerkraut Juice:*** Sauerkraut juice is produced in some countries for its dietetic value (lactic acid and vitamin C content) and its refreshing taste. The juice which is the result of the fermentation of lactic acid from cabbage, mainly from sliced sauerkraut, is used.

The juice must be the result of a normal lactic fermentation, *i.e.* without butyric fermentation or other deterioration.

A good quality juice must have an acidity of 1.4% lactic acid and a content of maximum 2.5% salt; this is obtained by the mixing of various sauerkraut qualities.

The collected juice (from sauerkraut production) is heated slightly in order to eliminate CO2 gas and to obtain protein coagulation. Filtration of juice is the next technological step, followed by filling in receptacles, closing of receptacles and pasteurization at 75-80° C for 4-5 minutes.

## CONCENTRATED TOMATO PRODUCTS

***Tomato Paste:*** The product with highest production volumes among concentrated products is tomato paste which is manufactured in a various range of concentrations, up to 44% refractometric extract. Tomato paste is the product obtained by removal of peel and seeds from tomatoes, followed by concentration of juice by evaporation under vacuum.

In some cases, in order to prolong production period, it may be advisable or possible to preserve crushed tomatoes with sulphur dioxide as described under semi-processed fruit "pulps".

Technological flow-sheets run according to equipment/ installation lay-outs, which are especially designed for this finished product. Manufacturing steps fall into three successive categories:

a. obtaining juice from raw materials;

b. juice concentration and

c. tomato paste pasteurization.

(a) Obtaining juice from raw material. Preliminary operations (pre-washing, washing and sorting / control) are carried out in the same conditions as for manufacturing of "drinking" tomato juice described above. Next operation is removal of seeds from raw tomatoes: tomato crushing and seed separation with a centrifugal separator.

Tomato pulp is pre-heated at 55-60° C and then passed to the equipment group for sieving: pulper, refiner and superrefiner with sieves of 1.5 mm, 0.8 mm and 0.4-0.5 mm respectively in order to give the smoothest possible consistency to the tomato paste.

(b) Juice is concentrated by vacuum evaporation, a technological step which in modern installations runs continuously, tomato paste from the last evaporation step being at the specified concentration.

In continuous installations with three evaporation steps (evaporating bodies), the juice is submitted in step / body I to pasteurization at 85-900 C for 15 min and this will determine the microbiological stability of finished product. Vacuum degree corresponding to this temperature is 330 mm Hg.

In evaporating bodies II and III, temperatures are around 42-46° C and vacuum at 680700 mm Hg.

Juice concentration occurs gradually and continuously in the three evaporating bodies.

The advantages of continuous concentration are as follows: the taste, colour, flavour, "shine" and consistency of tomato paste are improved because:

(i) the real concentration is performed in evaporating bodies II and III at low temperatures (42-46° C) and

(ii) the whole concentration process time from the input of juice in body I until the output of paste from body III is of about 1 hour (for paste with 30-35% refractometric extract).

- production capacity is raised by about 30% as compared to discontinuous installations with the same evaporation surface;
- the steam consumption is reduced by 60% because heating of bodies II and II is done with vapours

resulting from juice evaporation in body I (double effect); water and electricity consumptions are also reduced by 30-40%.

(c) Tomato paste pasteurization assures the microbiological stability of the product. For this purpose, the paste coming out from concentration equipment is passed continuously and in a "forced" mode through a tubular pasteurizer from which it emerge at a temperature of 90-92° C.

Usual commercial tomato paste types are at concentrations of 24%, 28% and 32% refractometric extract. Sometimes it is possible to obtain a tomato paste with a concentration of 44% refractometric extract; for this purpose it is necessary to eliminate a part of cellulose from tomatoes, an operation performed in a separating turbine. Tomato paste storage and preservation is carried out after packing which is done usually in drums, metallic cans or glass jars; some modern equipment has been developed for packing in aluminium bags. As far as the concentration of tomato paste is concerned it is not possible to reduce water content down to 30% which corresponds to a water activity aw of 0.700.75 (minimum limit of mould growing), it is necessary to take special measures (e. g pasteurization, cold storage or salt addition). Salt is not a preservative in itself but contributes to the lowering of water activity. In drums, the preservation of tomato paste with minimum 30% refractometric extract is carried out in two ways:

- the hot paste (about 90° C) flows directly from pasteurization equipment into drums that have been previously steamed;
- the paste is cooled down to 30° C through a heat exchanger and is introduced into drums that have been previously steamed.

For preservation purposes, it is possible to add 3-8% salt. Preservation with 3% salt must be carried out respecting the following criteria:

(a) processing of a healthy raw material;

(b) thorough washing and control;

(c) pasteurization of concentrated paste and use of well prepared drums. Paste in drums has to be stored in cold storage rooms during the hot season.

Preservation in big metal cans of 5 and 10 kg capacity of tomato paste with a minimum of 30% refractometric extract can be achieved without sterilisation if the following conditions are respected:

(a) sterilisation by steam of cans and covers;

(b) filling of paste at 92-94° C;

(c) airtight sealing/ closing of cans;

(d) invert cans and then

(e) air cooling.

For small packages (tinplate cans of 1/10-1/1 or glass jars of same capacity) it is usual to use pasteurized paste, as hot as possible (92-94° C). The receptacles are first sterilised by steam. After airtight sealing, the receptacles are kept in boiling water for a short time in order to sterilize their inner surface and the paste in contact with inner receptacle surface. In some countries small receptacles are not further sterilised if the manufacturing is carried out in perfect hygienic and sanitary conditions. Packing in small tinned aluminium tubes is carried out with concentrated paste, pasteurized and hot.

Good quality tomato paste is an homogenous mass, with a high density, without foreign bodies (seeds, peel, etc.), with a red colour, and an agreeable taste and smell, close to those of fresh tomatoes. There are usually three types of tomato paste: 36, 30 and 24 which have refractometric extracts of respectively 34-38%, 28-32% and 24-26%. Paste of good quality must have a volatile acidity of maximum 0.15% as lactic acid. An 8% salt addition is accepted.

***Concentrated Tomato Juice:*** Concentrated tomato juice is a product with 17-19 % refractometric extract and is a homogenous mass, finely sieved, without foreign bodies / and without any evidence of deterioration. A good quality product has a red colour, an agreeable and specific taste and smell. Modern technology uses the same installations, equipment and flow-sheets for concentrated tomato juice as for the production of tomato paste; the final concentration is thus regulated between the above specified limits. The concentrated tomato juice is filled in receptacles (metal tinplate cans or glass bottles) and then pasteurized at 100° C during 15-25 minutes according to receptacle type. With modern production lines it should be possible to pass the concentrated tomato juice through a tubular pasteurizer and then pack aseptically and cool, without the need to pasteurize the receptacles.

***Tomato Sauces:*** Under the USA Code of Federal Regulation 7 CFR 52, 1991 tomato sauce is the concentrated product prepared from the liquid extract from mature, sound, whole tomatoes, the sound residue from preparing such tomatoes for canning, or the residue from partial extraction of juice, or any combination of these ingredients, to which is added salt and spices and to which may be added one or more nutritive sweetening ingredients, a vinegar or vinegars, and onion, garlic, or other vegetable flavouring ingredients. The refractive index of the tomato sauce at 20° C is not $< 1.3461$. These products are widespread in some countries and are used in order to spice some meals. Sauces can be obtained from fresh tomatoes or from concentrated products (tomato paste or concentrated tomato juice), those from fresh tomatoes being of superior quality. Technological processing covers the following steps: concentrated juice processing, addition of flavour/taste ingredients (salt, sugar, vinegar, spices, etc.), boiling, fine sieving, filling of receptacles, closing and pasteurization (45 min at 85° C).

Tomato sauces which can be sweet, more or less spicy are prepared according to specific recipes. Production accidents and product defects; means to avoid them

*Tomato Juice:*

- "Separation" in layers is due to not enough homogenisation or low / insufficient viscosity. In the first case it is necessary to intensify homogenisation; and in second to increase the pre-heating temperature to 60° C in order to obtain protopectine hydrolisis and pectolitic enzymes inactivation.

- Moulding of the juice is brought about by significant infection of raw materials, inadequate washing and control or by use of contaminated packages. The preventive measures should be decided after cause analysis. Good pasteurization can destroy all moulds but the bad juice taste remains.
- Fermentation of juice is manifested by a significant development of gases. Prevention methods are the same as for moulding.
- Tomato juice turns sour, without the formation of gases; this defect is initiated by thermophyl and thermoresistant bacteria; the juice acquires a vinegary taste. Prevention: maintenance of flash pasteurization temperature at 130-135° C.
- Excessive vitamin C losses are due to a simultaneous action of heating and oxygen from air. Prevention:
  a. prevent air going into crusher and extractor;
  b. assure an intensive de-aeration (vacuum degree 700 mm Hg) at a temperature of at least 35-40° C; and
  c. close receptacles in vacuum.
    - Weak colour of tomato juice can be avoided by the utilisation of mature tomatoes and with a pulp of as red a colour as possible.

*Tomato Paste and Concentrated Juice:*

- Presence of sand is caused by inadequate washing or by a significant contamination of raw material; this can be prevented by a more intensive pre-washing and washing of tomatoes.
- There may be mould especially at the surface of tomato paste packed in drums. Prevention:
  a. accurate pre-washing and washing;
  b. follow pasteurization instructions;
  c. pack in clean drums or receptacles; and
  d. close receptacles immediately after filling.

- Fermentation is manifested by a weak alcohol smell or by a weak vinegar taste; when the fermentation is more advanced there is gas production in the product mass. Prevention: as for moulding prevention.

*Tomato Sauces:*

- Surface of the product turns black at the contact zone with air; this is due to the action of iron on the tannins from spices, tomato seeds, etc. Prevention:

  a. avoid iron equipment;

  b. avoid crushing of tomato seeds and

  c. seal receptacles in vacuum.

## Pickles and Sauerkraut Technology

*Vegetable Natural Acidification Technology:*

***Gherkins and Cucumbers:*** Raw materials must follow strict specifications for a high quality finished product; the following parameters must be considered as critical:

- adapt a uniform size according to the finished product requirements; for example, gherkins will need to have a maximum length of 9 cm for raw vegetables. Generally 15 cm size/length will be a maximum for high quality cucumber products in many countries. However, according to local preferences, bigger cucumbers could be also in demand.
- cylindrical or ovoidal shape;
- dark green colour;
- absence of surface defects due to cryptogamic diseases.

Cucumbers have to be picked at their ripeness for eating, when the sugar content is at about 1.5-2.2%, needed for lactic fermentation. Unripe cucumber does not have enough sugar.

The general technological flow-sheet is as follows:

*Reception*

*Control*

*Temporary storage*

*Grading by size*

*Washing*

Small holes are made in large size cucumbers skin;

*Receptacle filling: Raw material is simply put in the receptacles in bulk, with care to arrange them in such a way that a maximum of pieces could be introduced;*

*Salt solution preparation: 6% salt solution (nacl);*

*Salt solution addition: The salt solution is poured into the receptacle;*

Fermentation is carried out at 20-30° C, anaerobically. This step takes generally 4-8 weeks. Acidity reaches a value up to 1.5% lactic acid (and in some exceptional cases up to 2% lactic acid) which corresponds to a maximum pH value of 4.1.

Storage; after the last fermentation stage, drums and other receptacles have to be stored at low temperature; best conditions for 12 months shelf life should be below + 15° C. Storage temperature will determine the shelf life of the products. Addition of 1000 ppm potassium sorbate will prevent mould development without having any influence on lactic fermentation. Raw material grading by size is a very important technological step. In order to accelerate brine penetration, mainly for medium to large size cucumbers, the practice of making small holes in the raw material skins is generally recommended. A major factor influencing the quality of lactic fermented cucumbers is the water durity; optimal results are obtained at 15-20° durity.

Cucumber consistency/texture is influenced by the formation of calcium pectate with the pectic substances from raw material tissues. In some countries, calcium chloride (0.3-0.5 %) is added in order to firm up the cucumber consistency. Chlorinated water which still contains active chlorine can inhibit or even stop the lactic fermentation.

***Sauerkraut:*** In some countries cabbages are submitted to lactic fermentation as whole vegetables; however, in many countries the cabbage is shredded before fermentation. As shredded cabbage and its technology is at the basis of an important industry, giving good quality products, with a uniform fermented product and with good keeping quality and ease of distribution, this will be described first. Cabbage as raw material for sauerkraut must be sound, ripe for eating, well-leafed and from suitable varieties. Optimum total sugar level needed for the lactic fermentation is 24%; generally good quality raw material contains up to 30-60 mg/100 g of vitamin C.

***Shredded Sauerkraut:*** The technological flow sheet is as follows:

*Reception*

*Control*

Temporary Storage is carried out in bulk, up to a height of about 1 m, during few days. This step produces a heat generation which facilitates later fermentation by the softening of tissues.

**Removal of External Leaves**

Coring is done with a specially adapted mechanical screw; this operation generates small particles of finely divided cabbage which will be mixed with the main part of vegetable during shredding / chopping. The core represents about 10% from the whole cabbage, is rich in sugar and vitamin C, but being too high in fibre content needs to be chopped separately as described.

Shredding/Cutting of cabbage is carried out with complex specific equipment which is generally installed directly on the "top" of fermentation silos and is mobile, installed on rails and moves all along the silos. The dimension of resulting shredded cabbage is about 2-3 mm thick.

The same complex equipment is designed to grind the added salt to fine particles and to distribute shredded cabbage and ground salt in an uniform manner to the fermentation silos. The usual capacity of fermentation silos is up to 30 tons, with separate compartments of 45 tons each.

Salt Addition is carried out by the equipment described above; the proportion of salt is 2-2.5% with respect to cabbage.

This proportion must not be changed because the salt in this technology does not have a preservative role but only that to extract from cabbage the juice needed for fermentation. It is be preferable to obtain a fairly light pressure on cabbage just after salt addition with some simple mechanical means. This is important in order to:

- create an anaerobic medium for fermentation;
- facilitate external diffusion of cellular juice;
- assure a rational use of the fermentation space.

***Fermentation:*** The maximum acidity level obtained is generally of about 1.5% lactic acid (and very rarely 2.5%); this is obtained in 4-6 weeks. Optimal acidity is 1.0-1.8% and pH value 4.1 or lower. Fermentation temperature is at 20-25° C in the first phase and needs to be lowered then to 14-18° C. During fermentation, the brine from each storage / fermentation silo cell is periodically circulated with a pump in order to uniformise the fermentation process.

Storage is performed in same silos used for fermentation, or the finished products is removed from silos and packed in drums and other receptacles according to distribution schedule. These silos are usually made of reinforced concrete and coated with gritstone plates or with an acid-resisting material layer. At small scale and in traditional processing, shredded sauerkraut can be obtained by using simple available glass or rigid plastic receptacles. At home, this process can use glass jars and / or local / traditional pottery receptacles from a minimum size of 2-3 kg up to the available / practical sizes (better limited to 10-15 kg). In some countries shredded sauerkraut is preserved in receptacles by pasteurization, once the fermentation process has been completed.

***Whole Sauerkraut:*** According to the consumer preference in different countries and to the specific situations it is also usual to preserve whole cabbages by lactic fermentation. At small or medium scale operations, whole cabbage could be processed/ fermented in cylindric receptacles like 30 to 200 litre rigid plastic drums, or rectangular receptacles made from food grade rigid plastic. It is possible to find this type of drum in a significant number of developing countries. These two types of rigid plastic receptacles could also be used for shredded sauerkraut production. Prepared whole cabbages are put into fermentation receptacles and a 5-6 % salt concentration brine is poured on top. The fermentation conditions are the same as for shredded sauerkraut. In order to assure a uniform fermentation and to avoid a strict anaerobic (butyric) fermentation it is necessary to apply a periodic juice "aeration" (each 2-3 days at the beginning of the fermentation, and then each 5-7 days). A simple flow-sheet for preparation of whole sauerkraut at family / farm / community levels is presented in section 9.6.1.4.

***Other Acidified Vegetables:*** In principle all vegetables with a sugar content of at least 2 % could be preserved by lactic fermentation. From a practical point of view it is mainly the following vegetables which are preserved by this technology: unripe tomatoes (green tomatoes), peppers, eggplant, carrots and cauliflower, alone or usually in a mix with cucumber as mixed pickles. Fermentation of individual vegetables is carried out according to a flow-sheet as described for whole sauerkraut (section 9.6.1.4). The type of cut, brine concentration and frequency of operating steps have to be adapted to each case; green tomatoes are fermented as whole vegetable. Simplified flow-sheet for whole sauerkraut processing:

*Process:*

(a) Cabbage preparation:

- Remove the damaged leaves;
- Wash the vegetable;
- Remove 2-3 outer leaves;
- Size grade in three categories:
  - * size A: about 700 g per cabbage
  - * size B: less than 1.2 kg per cabbage
  - * size C: more than 1.2 kg per cabbage
- Process each size category separately;
- Wash cabbages;
- Remove cores;
- Cut size category C vegetable in halves.

(b) Salt solution (brine) preparation:

- Prepare a 5% salt (NaCl) solution = 500 g salt for 10 litre water or 50 g salt for 1 litre water;
- Stir until complete salt dissolution;
- Filter salt solution through cheese cloth.

(c) Initial processing:

- Use a different receptacle for each size category;
- Arrange cabbages in fermentation receptacle;
- Pour salt solution to completely cover cabbages;
- Fix some clean wood pieces (or better some fitted covers with holes) in order to keep cabbages completely covered by salt solution. Allow about 10 cm salt solution above cabbage level;
- Store fermentation receptacles in a moderately cold and ventilated place, out of direct sunlight / heat, protected from dust and other nuisances (insect, etc.);
- Cover each fermentation receptacle with a piece of cardboard or cloth.

(d) Processing follow-up:

- During the first week after initial processing. Once every 2 days, it is necessary to: remove the cover;
- collect and carefully remove carefully (with a household spoon) the white layer ("scum") formed at the surface of salt solution;
- wash the spoon each time and rinse;
- put back the cover.
- During the 5 following weeks.
- once every 4 days: repeat the operations described above.

After the first week, in order to assure a homogeneous acidification / fermentation process for big receptacles (*i.e.* drums or other receptacles of 20 to 2001 capacity), it will be necessary to proceed once a week to an "aeration" step. After completion of brine surface cleaning (as described above), the following operations will be carried out:

- remove the cover and the wood spacers.
- remove all salt solution (brine) from the receptacle;
- filter this solution through a cheese cloth;
- pour back the filtered solution back to the fermentation receptacle;
- put the wood spacers back in place.
- cover the receptacle.

These operations will be carried out for each fermentation receptacle once a week, during an estimated period of six weeks; total duration will be determined by the temperature in the storage room and by the chemical composition of specific raw material (cabbage) lots. Always keep salt solution (brine) level at 10 cm above cabbages, e.g. cabbages must be always covered by brine.

***Consumption of the Finished Product:*** It is possible to estimate that at reasonable ambient temperatures and with a strict followup of the above recommendations, the finished product will be ready for consumption about 6 weeks after initial processing. The finished product could be used "as is" in vegetable salads, or prepared according to local taste: with tomato sauce, beans, minced meat, etc. as a replacement of fresh cabbages.

In the same way as with natural acidification or lactic fermentation the cabbage texture is modified and softened so that tissues are more digestible than fresh vegetable. It is possible to use the finished product in local dishes and in new recipes without having to boil it. Apart from the taste benefits of acidified cabbages, this is also produces a significant fuel savings. The juice resulting from natural cabbage acidification is recovered and could be used separately as a refreshing vegetable juice; the preparation is described in this document.

***Finished Product Storage:*** It is possible to store the finished product after completion of fermentation (*i.e.* after the estimated six weeks period); the storage time will depend on the ambient air temperature. If a cool space is available, the finished product shelf-life/storage time at a temperature of about + 15° C is estimated at six months. At an ambient temperature not exceeding +20° C, the storage time could be estimated at 2-3 months.

## ARTIFICIAL VEGETABLE ACIDIFICATION TECHNOLOGY

This technology is based on the addition of food grade vinegar which has a bacteriostatic action in concentrations up to 4% acetic acid and bactericidal action in higher concentrations. Vegetables preserved in vinegar need to reach, after equilibrium between vinegar and water contained in vegetables, a final concentration of 2-3 % acetic acid in order to assure their preservation. To achieve this final concentration, a 6-9% acetic acid vinegar is used, as related to the specific ratios vinegar/ vegetables. In vinegar pickles, salt (2-3%) and sometimes sugar (2-5%) are also added.

If the vinegar concentration is lower than 2%, vinegar pickles need to be submitted to a pasteurization in order to assure their preservation.

***Cucumbers in Vinegar:*** This represents the basic product obtained by this technology. Cucumbers have to be wholesome, with a soft texture and not have reached eating maturity. They must have a low sugar content because in this technology there is no lactic fermentation involved. Dimensions are up to 12 cm length, with a preference for small cucumbers.

The technological steps are the followings:

*Size grading*

*Washing*

*Arrange in receptacles - glass jars, etc.*

Pouring of vinegar is usually carried out at room temperature; however, hot vinegar addition enables a sterilisation of cucumber surface and facilitates vinegar penetration in vegetable tissues.

*Salt (Sugar) Addition*

*Spicing Addition*

The technological cycle of artificial acidification is considered completed when acetic acid concentration reaches an equilibrium value; the time needed is about 2 weeks. When equilibrium concentration in acetic acid is below 2%, the cucumbers are submitted to a pasteurization for 20 min at 90-1000 C in order to assure their preservation. Cucumbers in vinegar with previous lactic fermentation are excellent quality products because the lactic fermentation improves the taste of these cucumbers. The principle of this process is to assure preservation both by acetic acid and by lactic acid simultaneously. Technological processing flow-sheet is as follows: small cucumbers ("cornichons" or "gherkins") are washed, brushed and small holes are made in the skin; the vegetables then are put in drums with slightly warm 6% brine which also contains spices.

The lactic fermentation runs for few days up to a lactic acid concentration of about 0.5 %. The cucumbers are removed from the brine, washed thoroughly and well drained. Preservation is usually done in glass jars by pouring a normally flavoured vinegar with about 9% acetic acid usually in order to bring the final concentration to 3% calculated as acetic acid.

In order to obtain a high quality product only wine vinegar should be used. In some pickles (*e.g.* in "Cornichons") the usual level of wine vinegar is set at 20 % of packaged product total weight; some alcohol vinegar could be still added and final concentration will be adjusted as described above.

***Other Vinegar Pickles:*** One type in this category is represented by other vegetables acidified with vinegar separately or in a mix (red peppers, sweet green pepper, green tomatoes, cauliflower, etc.). The preparation steps are similar to the ones used for cucumbers in vinegar.

Significant quantities of special mixed vegetables in vinegar are manufactured in many countries, with the international name of "mixed pickles" with following composition: small cucumbers ("cornichons"/"gherkins") - maximum 70 mm in length -, sliced carrots, cauliflower, small onions (less than 25 mm diameter), mushrooms etc. and spices.

The vegetables are acidified separately in vinegar and then are put into receptacles (glass jars); a flavoured vinegar, salted and sweetened with acetic acid concentration of 3-5% is poured over them. In the case of lower acetic acid concentrations, a pasteurization at 90° C for 10-20 minutes is applied according to the receptacle size.

## GENERAL HEAT PRESERVATION OPERATIONS—CANNING

The success of heat preservation operations lies in:

- selecting suitable fruit and vegetables in good conditions;
- preparing them hygienically and skilfully;
- packing them in cans which are hermetically sealed and then processed under fixed conditions of time and temperature;
- cooling these cans carefully and storing them under conditions which will not cause deterioration of either the cans or their contents.

### Selection of Raw Materials

It is appreciated that some varieties of fruit and vegetables are not suitable for canning, either because they are uneconomical to prepare or because the colour, flavour or texture are poor. Suitable varieties must be available to the canner in quantities sufficient to meet his requirements and in sound conditions for canning. The flow to the cannery should be regulated in order that perishable materials are not left for a long time before being handled, since any delay will cause deterioration.

Apart from the main ingredients, be it fruit or vegetables, minor ingredients also require careful selection. Sugar, salt, water and spices for instance may all be contaminated with spoilage organisms, so constant testing of all raw materials is essential.

***Preparation:*** This is carried out by various methods, including grading, trimming, peeling, washing and blanching. All equipment must be scrupulously clean and preparation should be completed quickly and carefully in order to keep the bacterial load as low as possible.

Thorough washing of vegetable is necessary to remove spores of heat resistant bacteria which are present in large numbers in the soil. Blanching in steam or hot water is of no avail against these heat resistant (thermophilic) spores because of the comparatively low temperatures involved. Reasons for blanching are:

- the removal of gas from the tissues of the raw material;
- the shrinkage of this material;
- the inhibition of enzymic reactions, which, if not checked, will adversely affect the colour and nutritive value of the food.

***Filling:*** Filling, be it mechanical or by hand, requires careful attention. The cans must be clean and the correct weight of foodstuffs must be added. Under-filled cans will be underweight and the headspace will be too large, resulting in too much air being left in the can. Overfilling may lead to seams being strained during processing and to ends becoming distorted and bulged.

If the product forms hydrogen on storage as is the case with coloured fruits, swelling of the can due to hydrogen pressure will occur more quickly in an overfilled can than in one which has been correctly filled. Overfilling also affects heat penetration in the can and may lead to spoilage outbreaks.

***Air Removal:*** Before the can is seamed, air must be removed from the contents and the headspace. Normally, this is carried out by passing the cans through a steam box until the temperature at the centre of the can is at least 160° F. This operation, termed exhausting, is necessary for the following reasons:

i. to minimise strains on the seams due to expansion of air during the processing period;

ii. to remove oxygen which accelerates corrosion in the can and also causes oxidation of the food with possible serious effects on colour and flavour;

iii. to reduce the destruction of vitamin C;

iv. to enable a vacuum to be formed when the can is cooled.

This ensures that the ends remain concave, even when storage temperatures are a little higher than usual, and also acts as a reservoir for hydrogen which may be formed by reactions between the can and its contents. Thus a high vacuum makes for a long shelf life.

Large cans, however, should not reach such a high exhaust temperature before seaming as smaller cans because of the danger of the can body collapsing on cooling, a condition known as "panelling".

***Double Seaming:*** The can should be double-seamed as soon as the correct centre temperature has been attained. Any delay between exhausting and seaming will lead to loss of vacuum and may lead to bacterial spoilage. The quality of the double seam must, of curse, be frequently checked.

***Heat Processing:*** After seaming, the cans are heated for a definite time at a definite temperature to kill or inhibit organisms which may cause spoilage. This operation is termed "heat processing".

The times and temperatures required for "heat processing" of various packs have been determined experimentally to ensure that spores of the most heat resistant food poisoning organisms known, Clostridium botulinum, are destroyed. There are other organisms, however, whose spores are more heat resistant than those of Clostridium botulinum and which although they will not cause food poisoning may cause spoilage and for this reason the minimum heat processing time is often exceeded by recommendations made by laboratories.

At the same time there is a limit to the amount of heating which a canned food may be given without spoiling its flavour, texture and colour and this also has to be taken into consideration when process recommendations are made.

Bacterial spores have a greater resistance to heat when the growth-medium is neutral or near neutral, and neutrality is normally required for bacterial growth to commence. Because of this, canned foods have been broadly divided into two groups:

(a) "acid" foods having a pH of 4.5 or lower and

(b) "non-acid" foods having a pH of more than 4.5.

"Non-acid" foods (vegetables) must, therefore be "heat processed" at high temperatures using steam under pressure, whereas "acid" foods (fruit) may be processed at the (lower) temperature of boiling water, since this will kill moulds and yeasts and if any bacterial spores survive the combination of acid and heat, they will be inhibited from growth by the acid environment.

The rate of destruction by heat follows a definite pattern, the same proportion of the surviving bacteria being destroyed in successive units of time.

The more bacteria there are in a pack, the more time will be need to reduce their numbers. For this reason, it is essential that the initial number of bacteria be kept low, and this may be achieved by ensuring fast and hygienic handling at all stages in the cannery.

Pressure gauges and retort temperature control equipment must be checked frequently for accuracy. Processing times and temperatures must be strictly adhered to, and complete removal of air from the retort during processing must be achieved by adequate venting. Failure to remove the air completely will result in their being cold spots in the retort and intermittent spoilage is likely.

***Cooling:*** As soon as the heat processing time is completed, the cans are cooled in chlorinated water as rapidly as possible without damaging them.

Cans processed in steam develop high internal pressure because of the expansion of the foodstuff, the expansion of air in the can and the increase in the vapour pressure of the water in the can. During the heat process, these pressures are counterbalanced to some extent by the pressure of the steam in the retort, but on releasing this steam pressure at the commencement of the cooling period, the pressure in the can may be sufficient to strain the seams seriously and may even distort the ends.

Cans of A21/2 size or larger, when processed at temperatures of 240° F or more, are liable to undergo permanent distortion, such as peaking. This may be avoided by pressurecooling, which involves replacing steam pressure by air pressure before introducing water to the retort, and maintaining this until the pressure inside the can has fallen to a safe level.

This presents difficulties, since if the air pressure is maintained after the can has developed a vacuum, the can body is liable to collapse. Where pressure-cooling is not carried out, the retort pressure is allowed to drop slowly to atmospheric pressure and the cans are then cooled with water.

***Storage:*** After cooling, the cans should be stored in cool, dry conditions. The maintenance of a constant temperature is desirable, since a rise in temperature may lead to condensation of moisture on the can, with possible rusting.

Cool conditions are required because storage at higher temperatures not only causes chemical and physical changes in the product and the container but also introduces a risk of thermophilic spoilage. Other known causes of container spoilage in storage are the use of labels and cardboard cases which have too high a chloride content, and the use of unseasoned wood in the manufacture of packing cases, all of which tend to cause rust formation on the cans. General technical operations for fruit and vegetable canning lines:

(a) Receptacle washing will remove the impurities and, as much as possible, the microorganisms on the inner surface of metallic cans or glass jars. Washing must be performed just before receptacle filling in order to avoid a new contamination.

Washing methods are variable and depend on receptacle type and need to be carried out with adequate mechanical equipment.

Metal cans are washed on the can feeding lines of filling equipment; a high pressure spray of warm water (65-8° C) is directed into the receptacles while these are submitted simultaneously to a rotation and forward motion.

Glass jars are submitted to a triple washing: wetting for 10 min in a warm detergent and disinfectant solution (40-45° C) containing 100 mg active Cl/litre; washing with high pressure (2.5-3 at) warm water sprays (65-85° C); rinse with cold water. Special attention MUST be given to recycled glass jars; washing process must be intensified or repeated, depending upon their contamination.

(b) Receptacles are filled in order to maintain a specific ratio between the solid part of the composition and the filling or covering liquid.

For canned vegetable products, the covering liquid may be a 1-3% salt solution with or without addition of sugar (1-3%), tomato concentrated juice or various sauces based on concentrated tomato juices.

Salt solution (brine) preparation may be performed with salt percolators; the resulting solution is saturated, containing 318 g/l and needs to be diluted to usual concentrations (1-3%). Brine is then heated up to filling temperatures which depend on product type (up to 85-90° C).

Sugar solutions (syrups) for fruit products may be prepared on the same type of percolators as brine.

Receptacle filling is carried out by leaving an empty space of 5-15% of the total volume, depending on filling temperature and the product type.

(c) Pre-heating (exhausting) of full receptacles aims at the removal of air from the tissues and the increase of the initial temperature of the receptacle contents. On modern production lines, exhausting is eliminated and replaced by the increase of the filling liquid temperature and hermetic receptacle closing under vacuum.

When exhausting is applied, with steam or with hot water, the pre-heated receptacles must be immediately closed in order to avoid the contraction of liquid phase and thus air introduction. Exhausting is performed in special, continuous equipment; product temperature is between 80 and 95° C, during 2-10 min.

## QUALITY CONTROL/QUALITY ASSURANCE

The international trade in processed fruits and vegetables is very large with an ever increasing number of different types being processed and exported. Whereas once, processing was limited to mostly temperate climate fruits and vegetables, the change has now broadened to include tropical and subtropical types.

The reasons are twofold. Firstly, consumers' dietary habits have become more diverse so that, for example people living in North America may very well like fruit and vegetables grown in Africa or Asia. Secondly, processing techniques, whether they be for canning, freezing or drying, have been improved to an extent where final product is palatable, nutritious and of long and reliable shelf life.

Many developing countries have taken advantage of the continuing worldwide demand for processed fruits and vegetables and earned valuable foreign exchange from exports of products to profitable markets.

The export quality control and inspection of processed fruits and vegetables is directed at ensuring that the final products:

have been processed in a registered export establishment that is constructed, equipped and operated in an hygienic and efficient manner; conform to the requirements of the export regulations for processed fruits and vegetables, and those of the importing country, in respect of such things as quality grades, defects, ingredients, packaging materials, styles, additives, contaminants, fill of container, drained weight; and conform to labelling requirements.

### Inspection and Certification Procedures

In most countries, in processing fruits and vegetables for export, it is not customary to apply continuous inspection as it is in the case of meat. Few, if any, importing countries require it, and the nature of the products themselves is such that only part time check inspection is required during processing together with statistically based inspection, including sampling and analysis, of final product.

However, in circumstances where an establishment is processing export product for the first time, it can be argued that there is an advantage in adopting continuous inspection until the operation is satisfactorily established.

In any event, inspection of raw materials should be carried out at the commencement of each processing run to ensure that only sound fruit or vegetables of sufficient maturity (degree of ripeness) are used for processing. Sampling checks of raw materials should be carried out as frequently as the inspector thinks necessary.

The inspector must ensure that adequate hygiene practices are followed during the processing of the product. For example, in the case of canned and frozen products and other processing methods, raw materials should be washed absolutely clean so that fruit and vegetables entering the processing line are free from dirt, superficial residues of agricultural chemicals, insects and extraneous plant material.

In the case of dried product, especially where the raw material is sun-dried on drying greens or racks, care must be taken to minimize contamination by bird and animal droppings, dust and extraneous plant material. It is often necessary to wash the dried product to ensure cleanliness of the final product.

In the case of canning and freezing, the inspector must obtain full details of the processing programme for at least the following day from management, so that an adequate inspection programme can be scheduled.

In much the same way as for fresh fruit and vegetables, the inspector must also be aware of the pesticides and other chemicals used in the production of the raw materials. Necessary laboratory analyses can then be arranged to ensure residue levels in the final product do not exceed tolerances adopted by importing countries.

At the commencement of and during processing, the inspector should pay attention to the state of raw materials, the preparation of raw materials for processing (peeling, slicing, dicing, blanching, etc.), preparation and density of packing medium (sugar syrup, salt brine, etc.), the state of cans or containers to be used (cleanliness and strength), the cooking or freezing process (time/temperature relationship), can filling and closure and can/container storage.

After processing, the inspector should check the final product to ensure the drained or thawed weight, the vacuum and headspace, packing medium strength and that can/container conditions are satisfactory. Statistically based sampling plans should be adopted for the examination of final product to ensure it meets the requirements of the export regulations. The labelling applied to cans/containers should also be checked to ensure both their correctness and compliance with the export regulations and the requirements of those countries in which the product is to be marketed.

Cans should also be examined to make sure that the correct embossing relating to the product, its date of production and the registered number of the export establishment has been applied. Each establishment registered for the export of processed fruit and vegetables or for canned or frozen foods should have its own quality laboratory sufficiently equipped and staffed to carry out physical, chemical and microbiological examinations of the goods.

Inspectors should have access to the laboratory facilities and the establishment's quality control records as and when required. Independent laboratory examination of product should be made by the agency having responsibility for export on the basis of a statistically developed sampling plan.

In those countries where fruit and vegetable production is a seasonal event, processing for export generally takes place at the time of peak production and then declines, often to a halt, as the supply of raw materials declines. As a result, most export establishments produce at their peak of production far more product than they export at that time.

Therefore, most manufacturers find it necessary to store product for considerable periods before it is exported. Thus, proper storage is essential if the product is to retain its quality and cans remain untarnished.

Inspectors should regularly inspect storage facilities, noting their conditions and that of the stored product, looking for signs of deterioration such as pest infestation and rusting of cans. Prior to export, the exporter should be required to notify the export quality and inspection agency of his intention to export in accordance with the provisions of the export processed fruits and vegetables regulations and on the prescribed "Notice of Intention to Export" form.

The notice should be submitted in sufficient time before the shipment date to enable the product to be inspected satisfactorily; the intensity of inspection depending on the original state of the product, the conditions under which it has been stored and the length of storage. When product is approved, the agency will issue the exporter an "Export Permit" authorizing Custom's clearance of the product.

***Labelling:*** Customers and consumers expect the labelling on food to be a true description of what they are buying.

Misleading or fraudulent labelling is an unfair trade practice that cannot be tolerated. Most countries now have labelling laws stipulating how foods are to be labelled and what information labels must contain. Most, if not all of those laws have in common requirement that the label should bear:

- A statement of identity and a true, as distinct from misleading, description of the product;
- A declaration of net contents (weight or number of pieces);
- The name and address of the manufacturer, packer, distributor or consignee, and
- A list of ingredients (in descending order of volume or weight).

In addition, labels may also be required to include, amongst other things, the country of origin, date of manufacture or packing, a use-by or expiry date, nutritional qualities or values of the food, storage directions, a quality grade and directions for preparing the food.

More frequently than is often realized, consignments of food exports arriving on foreign markets are not permitted entry because the labelling does not comply with the mandatory requirements of the importing country.

This sometimes results in consignments being rejected, but more often in them being withheld from entry until the labelling is corrected or new labelling applied. In either case, trade is interrupted and the cost involved may make sales unprofitable. It is essential therefore, that exporters be familiar with the food labelling requirements of importing countries.

# BIBLIOGRAPHY


Brandwein, P.F.: *Sourcebook for the Biological Sciences*, San Diego, Harcourt Brace JOvanovich, 1986.

Chrispeels, Maarten : *Plants, Genes and Crop Biotechnology*, Sudbury MA, Jones and Barlett Publishers, 2003.

David Sadava: *Plants, Genes and Crop Biotechnology*, Sudbury MA, Jones and Barlett Publishers, 2003.

Fransman M, Junne G, Roobeek A: *The Biotechnology Revolution?*, Oxford, Blackwell, 1995.

George S. Paul: *Beyond Humanity: Cyber Evolution and Future Minds*, Roackland, Charles River Media, 1996.

Goodsell, David S.: *Bionanotechnology: Lessons From Nature*, Hoboken, Wiley-Liss, 2004.

Moravec, Hans: *Mind Children: The Future of Robot and Human Intelligence*, Cambridge, Harvard University Press, 1988.

Murray, David: *Seeds of Concern: The Genetic Manipulation of Plants*, Sydney, University of New South Wales, 2003.

Postman, Neil: *Technopoly: The Surrender of Culture to Technology*, New York, Vintage Books, 1992.

Retzer, W.J.: *Biotechnology Workbook*, Englewood Cliffs, Prentice Hall, 1991.

Shetty, Kalidas: *Food Biotechnology*, New York, Dekker/CRC Press, 2005.

Smith, John E.: *Biotechnology*, Cambridge, Cambridge University Press, 2004.

Towle, Albert: *Modern Biology*, Austin, Holt, Rinehart and Winston, 1988.

Walden, Richard: *Genetic Transformation in Plants*, England, Open University Press, 1988.

# INDEX

□□□